T0214017

SpringerBriefs in Applied Sciences and Technology

PoliMI SpringerBriefs

Editorial Board

Barbara Pernici, Politecnico di Milano, Milano, Italy

Stefano Della Torre, Politecnico di Milano, Milano, Italy

Bianca M. Colosimo, Politecnico di Milano, Milano, Italy

Tiziano Faravelli, Politecnico di Milano, Milano, Italy

Roberto Paolucci, Politecnico di Milano, Milano, Italy

Silvia Piardi, Politecnico di Milano, Milano, Italy

More information about this subseries at http://www.springer.com/series/11159
http://www.polimi.it

Paolo Rosa · Sergio Terzi
Editors

New Business Models for the Reuse of Secondary Resources from WEEEs

The FENIX Project

POLITECNICO
MILANO 1863

Editors
Paolo Rosa
DIG
Politecnico di Milano
Milan, Italy

Sergio Terzi
DIG
Politecnico di Milano
Milan, Italy

ISSN 2191-530X ISSN 2191-5318 (electronic)
SpringerBriefs in Applied Sciences and Technology
ISSN 2282-2577 ISSN 2282-2585 (electronic)
PoliMI SpringerBriefs
ISBN 978-3-030-74885-2 ISBN 978-3-030-74886-9 (eBook)
https://doi.org/10.1007/978-3-030-74886-9

© The Editor(s) (if applicable) and The Author(s) 2021. This book is an open access publication.
Open Access This book is licensed under the terms of the Creative Commons Attribution 4.0 International License (http://creativecommons.org/licenses/by/4.0/), which permits use, sharing, adaptation, distribution and reproduction in any medium or format, as long as you give appropriate credit to the original author(s) and the source, provide a link to the Creative Commons license and indicate if changes were made.

The images or other third party material in this book are included in the book's Creative Commons license, unless indicated otherwise in a credit line to the material. If material is not included in the book's Creative Commons license and your intended use is not permitted by statutory regulation or exceeds the permitted use, you will need to obtain permission directly from the copyright holder.

The use of general descriptive names, registered names, trademarks, service marks, etc. in this publication does not imply, even in the absence of a specific statement, that such names are exempt from the relevant protective laws and regulations and therefore free for general use.

The publisher, the authors and the editors are safe to assume that the advice and information in this book are believed to be true and accurate at the date of publication. Neither the publisher nor the authors or the editors give a warranty, expressed or implied, with respect to the material contained herein or for any errors or omissions that may have been made. The publisher remains neutral with regard to jurisdictional claims in published maps and institutional affiliations.

This Springer imprint is published by the registered company Springer Nature Switzerland AG
The registered company address is: Gewerbestrasse 11, 6330 Cham, Switzerland

Preface

This book summarizes the work being undertaken within the FENIX European Research Project (Grant Agreement N°: 760792, H2020-NMBP-22-2017—Business models and industrial strategies supporting novel supply chains for innovative product-services). The project aims at addressing the paradigm shift from linear to circular economy through a holistic perspective, where a set of circular business models and supply chains are demonstrated in practice in order to enhance potential benefits coming from the adoption of circular practices. FENIX demonstrates and validates its circular business models on three new supply chains able to link actors from different industrial sectors under the same umbrella and having the same intent to create new, high-value, product-services starting from e-wastes. All the three business models share the same waste stream as input material and same pilot plants. Moreover, FENIX assesses, studies and validates an integrated life cycle analysis (LCA) and life cycle cost (LCC) strategy, to guarantee the sustainability of the proposed business models.

The book covers many topics, since FENIX really involved several issues related with circular economy, industry 4.0 and product-service systems.

Chapter 1 introduces the main research contents and provides an overview on FENIX objectives.

Chapter 2 focuses on circular business models (CBMs). It describes the whole process implemented by FENIX partners to assess, identify, select and adopt the CBMs to be validated and demonstrated within the project. It also shows the main results from each stage.

Chapter 3 describes the circular economy performance assessment (CEPA) methodology. This methodology will be adopted to quantify the circularity level of each business model selected in FENIX.

Chapter 4 presents a semi-automated assembly–disassembly pilot plant. This pilot has been exploited in FENIX to disassemble both smartphone-shaped toys and real printed circuit boards (PCBs) extracted from smartphones.

Chapter 5 describes a mobile pilot plant for the recovery of precious and critical raw materials from e-wastes. This pilot has been exploited in FENIX to recover secondary raw materials from PCBs through a hydrometallurgical process.

Chapter 6 focuses on an innovative direct ink writing (DIW)-based additive manufacturing process. This process has been exploited in FENIX to reuse secondary materials from e-wastes as green materials for additive manufacturing processes.

Chapter 7 relates with a methodology aimed at carrying out the life cycle performance assessment of selected business models. For each of them, the methodology assesses economic, environmental and social impacts related with identified circular supply chains.

Chapter 8 presents a decision-support system for the digitization of circular supply chains. This system will allow a real-time interconnection among all the pilot plants involved in FENIX, by enabling a remote monitoring of their states.

Chapter 9 describes the user participation and social integration through ICT technologies. In order to involve both private and industrial end users, a dedicated marketplace and a forum have been developed in FENIX. They will allow a direct relation and involvement with customers.

Chapter 10 relates with the validation of the three new supply chains developed in FENIX. Here, training activities, prototyping and small-series production of finished products will be described into detail.

Milan, Italy   Paolo Rosa
                                                                                               Sergio Terzi

Contents

Chapter 1
Introduction

Paolo Rosa, Sergio Terzi, and Bernd Kopacek

Abstract This chapter aims at clarifying the main research contents and presenting the main objective of the FENIX project. we briefly describe some fundamental concepts, like Circular Economy (CE), Industry 4.0 (I4.0) and Product-Service Systems (PSSs). All these contents are strictly connected in FENIX, since the project aims at demonstrating the benefits coming from the adoption of CE-related practices through a set of PSS-based business models supported by I4.0-based technologies.

1.1 Circular Economy

Commonly agreed definitions of CE are those proposed by [1, 2]. First, the CE is defined as a global economic model to minimize the consumption of finite resources, by focusing on intelligent design of materials, products, and systems. Second, the CE aims at overcoming the dominant linear (e.g., take, make, and dispose) economy model (i.e., a traditional open-ended economy model developed with no built-in tendency to recycle; [3, 4]). However, only in the last few years has the relevance of the CE been amplified worldwide [5]. Before the CE was introduced, a traditional (linear) lifecycle was the only process followed during the conceptualization, design, development, use, and disposal of products [2]. Progressively, closed-loop patterns—completely focused on balancing economic, environmental, and societal impacts—have substituted old industrial practices.

P. Rosa (✉) · S. Terzi
Department of Management, Economics and Industrial Engineering, Politecnico di Milano, Piazza L. da Vinci 32, 20133 Milan, Italy
e-mail: paolo1.rosa@polimi.it

S. Terzi
e-mail: sergio.terzi@polimi.it

B. Kopacek
SAT—Austrian Society for Systems Engineering and Automation, Beckmanngasse 51/28, 1140 Vienna, Austria
e-mail: bernd.kopacek@sat-research.at

© The Author(s) 2021
P. Rosa and S. Terzi (eds.), *New Business Models for the Reuse of Secondary Resources from WEEEs*, PoliMI SpringerBriefs,
https://doi.org/10.1007/978-3-030-74886-9_1

1.2 Industry 4.0

Differently from CE, there is no consensus among experts about which technologies can be classified under the I4.0 umbrella. Thus, we decided to follow an alternative strategy during the implementation of this work. Initially, they adopted the nine pillars described by [6] as keywords to exploit during the literature assessment. Basing on the resulting literature gathered from the web, only five of the nine pillars were further assessed. This way, cyber-physical systems (CPSs), the IoT, big data and analytics (BDA), additive manufacturing (AM), and simulation were identified as the main I4.0-based technologies related to the CE. For clarification, brief descriptions of these four technologies are provided. First, CPSs are an integration of computation and physical processes. Embedded computers and networks monitor and control the physical process, usually with feedback loops, where physical processes affect computations, and vice versa [7]. Second, the IoT are technologies that allow interaction and cooperation among people, devices, things, or objects through the use of modern wireless telecommunications, such as radio frequency identification (RFID), sensors, tags, actuators, and mobile phones [8]. Third, BDA is the application of advanced data analysis techniques for managing big datasets [9]. Fourth, AM describes a suite of technologies that allow the production of a growing spectrum of goods via the layering or 3D printing of materials [10]. Finally, simulations consider a wide range of mathematical programming techniques to achieve purposes related to CE and I4.0 paradigms. What is rarely assessed by the literature is the relation between I4.0 and the CE, and their reciprocal effect on the overall performance of a company.

1.3 Product-Service Systems

The adoption of the service business by manufacturing companies is a common trend in many industrial sectors, especially those offering durable goods. This shift, referred to in literature as servitization process, is defined as "[…] the increased offering of fuller market packages or 'bundles' of customer focused combinations of goods, services, support, self-service and knowledge in order to add value to core product offerings" [6]. Servitization supports companies to strengthen their competitive position thanks to the financial, marketing and strategic benefits led by the integration of services in the companies' offer [6–9]. Differentiation against competitors, hindering competitors to offer similar product-service bundles and the increasing of customer loyalty are the main benefits of servitization. Today, more than ever, servitization is customer driven [10]. A research field that is often associated to the servitization process is the one related to the Product Service-Systems (PSS) [11]. The first definition of a PSS was given in 1999: "A product service-system is a system of products, services, networks of players and supporting infrastructure that continuously strives to be competitive, satisfy customer needs and have a lower environmental impact than traditional business models" [12]. Manzini points out that PSS is an innovation strategy that allows fulfilling specific customer needs [13]. Tukker observes that PSS is capable to enhance customer loyalty and build unique relationships since it follows customer needs better [14]. Another important contribution comes from Sakao and Shimomura that see PSS as a social system that enhances social and economic values for stakeholders [15]. The move towards the PSS entails an organizational change that makes a company shift from a product-oriented culture to a service-oriented one. The transition is quite a complex process that requires several changes and that usually happens in subsequent steps. Martinez et al. identify the five categories of challenges a company must deal with when moving along the servitization process, namely embedded product-service culture, delivery of integrated offering, internal processes and capabilities, strategic alignment and supplier relationships [16]. PSS often include value adding services based on ICT contributions, both in terms of enhanced information and knowledge generation/sharing, as well as of additional functionalities [17, 18]. PSS providers need to establish collaboration among specialized companies. Fisher et al. discussed approaches for service business development on a global scale. They consider organizational elements, such as customer proximity or behavioural orientation [19]. The closer affiliation of customers and manufacturers/service providers offer potential to generate revenue throughout the entire lifecycle [18, 20]. Moreover, as stated by Baines et al., "... integrated product-service offerings are distinctive, long-lived, and easier to defend from competition based in lower cost economies ..." [18]. The potential extension of the lifetime of tangible components of PSS, due to their integration with adding value services, opens interesting perspectives also about environmental sustainability improvements. The advantages coming from PSS have been demonstrated in literature, yet for many companies efficiently managing the service operations is still a challenge. Best practices and empirical analysis are mainly carried out with a focus on larger companies. Nonetheless, the PSS topic is more and more recognized by SMEs that are looking for innovative business solutions to improve their competitive advantages.

1.4 The FENIX Project

Since the advent of globalization, the European manufacturing sector is coping with both an increasing lack of stability in the market and a need for quicker responses to customers' demands. With time, these two elements disincentivated long-term investments of companies in tangible fixed assets, by shifting their attention in high-value markets characterized by lower volumes than mass production. Subsequently, plant's capacity use rates have felt down quickly since the production was moved abroad. This negative scenario has affected the overall performances of SMEs. In parallel, in Europe there has been an increasing awareness about the environmental impact of products and processes and the importance of the sustainable use of resources. In this context, the circular economy paradigm is getting more and more success.

The main aim of FENIX is the development of new business models and industrial strategies for three novel supply chains to enable value-added product-services:

- A modular, multi-material and reconfigurable pilot plant producing 3D printing metal powders. This pilot plant will allow the production of high-quality metal and CerMet powders to be used in the production of mechanical components through manufacturing processes like additive manufacturing (SLM, LMD) thermal spraying and sintering. The peculiarity of this use case is that the metallic material entering the manufacturing process will be recovered from different kinds of wastes coming from the mass electronics sector. These wastes, once disassembled to recover hazardous components, will be reduced in powders. Subsequently, powders will be separated in metal and non-metal ones. In this case, only some specific metals (e.g. Sn, Ni, Cu, Co and Al) present in powders will be refined completely through bio-hydrometallurgical processes, processed by High Energy Ball Milling and optimized by classification and jet-mills to be used in industrial 3D printing, thermal spraying or sintering processes.
- A modular, multi-material and reconfigurable pilot plant producing customized jewels. This pilot plant will allow the production of customized jewels through additive manufacturing processes. The peculiarity of this use case is that precious metals entering the additive manufacturing process will be recovered from different kinds of waste coming from the mass electronics sectors. These wastes, once disassembled to recover hazardous components, will be reduced in powders. Subsequently, powders will be separated in metal and non-metal ones. In this case, only precious metals (e.g. Au, Ag, Pt and Pd) present in powders will be refined completely through bio-hydrometallurgical processes and directly used as basic material in dedicated 3D printing processes.
- A modular, multi-material and reconfigurable pilot plant producing 3D printing advanced filaments. This pilot plant will allow the production of advanced filaments through additive manufacturing processes. The peculiarity of this use case is that both metals (e.g. Cu and Al) and non-metal resins entering the additive manufacturing process will be recovered from different kinds of waste coming from the mass electronics sectors. These wastes, once disassembled to recover hazardous components, will be reduced in powders. Subsequently, powders will be separated in metal and non-metal ones. In this case, only Cu, Al and a specific set of non-metal materials (e.g. ABS and epoxy resins) present in powders will be refined completely through bio-hydrometallurgical processes and directly used as basic material in dedicated 3D printing processes.

All the three pilot plants will share the same structure. They will be designed also to host and fully exploit industry 4.0 solutions represented by smart sensors able to send real-time data to the online marketplace developed in FENIX. This will enhance the sharing of overcapacity among different supply chains from very different sectors, the involvement of private end users in industrial processes as well as the provision of new services to companies, for the monitoring and control of the pilot plant.

The second aim of FENIX is the representation of a set of success stories coming from the application of circular economy principles in different industrial sectors. FENIX will demonstrate how the adoption of circular economy principles can enable more sustainable supply chains, by increasing quality, market value and alternative exploitation of secondary materials. FENIX will enable a concrete sharing of capacity among different industrial contexts and the active participation of local communities in industrial processes. This will enable a long-lasting European leadership in innovative manufacturing plants engineering.

The design and engineering needed for the three pilot plants will follow a similar logic. All the pilot plants must be modular, focused on multi-materials and easily reconfigurable. These three features are the basic enablers for the adoption of the same pilot in different industrial contexts. From one side, modularity will allow: (a) the selection of the only set of modules constituting the overall FENIX pilot needed by end users and (b) the parallel use of each FENIX pilot plant's module for different purposes. For example, a user interested only in recovering materials from wastes will decide to exploit the only assembly/disassembly and materials recovery modules, without considering the additive manufacturing one. From another side, the focus on multi-materials could guarantee a wider exploitation of the FENIX pilot— or better its materials recovery module—for treating several kinds of wastes, even different from the ones selected during the project. From the last side, the easiness of reconfiguration could allow a shorter setup time for adequating the FENIX pilot to different recovery/production processes.

All the FENIX pilot plant's modules are based on already existing pilot plants:

- Industry 4.0 Lab: POLIMI is going to implement within its Department of Management, Economics and Industrial Engineering a pilot plant dedicated to assembly/disassembly activities. This demonstrative, lab-scaled, manufacturing process will be adequately reconfigured to manage the selected kind of obsolete products that could be the source of materials to be recovered during FENIX.
- HydroWEEE/Demo pilot plant: UNIVAQ, together with SAT and GREEN, has already implemented a mobile pilot plant dedicated to the recovery of materials from electronic wastes. This chemical process will be adequately reconfigured to manage different kinds of materials in a more sustainable way.
- High Energy Ball Milling pilot plant: MBN has already developed a pilot plant for High Energy Ball Milling of metal and ceramic materials producing powders for additive manufacturing and thermal spraying purposes. This pilot plant will be adequately reconfigured basing on the new requirements of the FENIX project.

The third aim of FENIX is the integration of Key Enabling Technologies (KETs) for the efficient recovery of secondary resources. FENIX will support the integration of different KETs within a unique industrial plant. Industry 4.0 and circular economy principles will be considered in the project, in order to enhance the development of innovative business models and supply chains based on new kinds of product-service concepts. Essentially, two types of KETs will be considered by FENIX:

1. Advanced manufacturing systems: a wide number of sensors will be embedded within each module constituting the FENIX pilot plant. These components will have a double role. From one side, they will allow a real-time control and optimization of operational procedures. From another side, they will allow the real-time sharing of information with the society. At the same time, the integration of automated assembly/disassembly procedures, advanced materials recovery techniques, additive manufacturing technologies and the digital world will put together sustainable processes and local societies.
2. Industrial biotechnology: since the initial steps, FENIX considered the exploitation of biometallurgy for the sustainable recovery of materials from different kinds of wastes. The final aim is demonstrating not only the environmental and social sustainability related to this type of processes, but especially their economic relevance.
3. Nanotechnology: this kind of material technologies enables an improvement of mechanical properties of materials, as well as thermal and electrical conductivity and functional properties. These technologies open the market to new materials able to substitute the most critical ones used today and seek more lightweight solutions than current materials. The High Energy Ball Milling process will induce a nano-structurization in materials that will be retained in the manufacturing process by additive laser sintering, thermal spraying and fast sintering.

These three aims are represented all together in the following Fig. 1.1, which represents how circular economy principles and digital tools will be used in the project to implement and test the three different modules constituting the FENIX pilot plant, using an iterative flow of data and knowledge between different actors involved in the supply chain. FENIX will allow closing the material's loop between original production, usage and final recovery, providing IT tools supporting a continuous cooperation between industrial and private contexts, from one side, and different industrial sectors from the other side. This approach will help advanced materials recovery techniques to reach better new market requirements, sharing overcapacity and better linking industrial plants with local communities, by also increasing the European manufacturing sector competitiveness worldwide. FENIX, through a wide usage of sensors and social media, will collect information from the plant and will share them with different end users, supporting them in several daily operational aspects.

When products will reach their end-of-life, becoming waste, they will be collected and sent to the FENIX pilot plant. Here, the manufacturing/demanufacturing module will disassemble them, by extracting only the most relevant components (basing on materials contents). These components will be shredded and reduced into powders. Once separated basing on their physical characteristics, the biometallurgical module will recover and refine metals. The additive manufacturing module, for producing value-added products, will exploit these metals (see use case description for details). In case of complex products being produced, the manufacturing/demanufacturing module will be reconfigured for managing and automate the final assembly process.

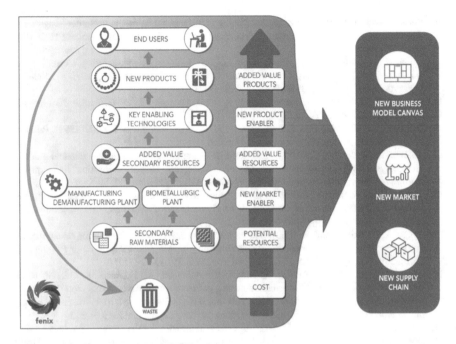

Fig. 1.1 Overall concept of the FENIX project

References

1. The Ellen MacArthur Foundation. (2015). *Towards a circular economy: Business rationale for an accelerated transition.*
2. Su, B., Heshmati, A., Geng, Y., & Yu, X. (2013). A review of the circular economy in China: Moving from rhetoric to implementation. *Journal of Cleaner Production, 42,* 215–227. https://doi.org/10.1016/j.jclepro.2012.11.020.
3. Stahel, W. R., & Reday-Mulvey, G. (1981). *Jobs for tomorrow: The potential for substituting manpower for energy.* (1st ed.). Vantage Press.
4. Pearce, D. W., & Turner, R. K. (1991). Economics of natural resources and the environment. *Land Economy, 67*(2), 272–276. https://doi.org/10.1016/0308-521X(91)90051-B.
5. Reuter, M. A., Hudson, C., van Schaik, A., Heiskanen, K., Meskers, C., & Hagelüken, C. (2013). *Metal recycling: Opportunities, limits, infrastructure, a report of the working group on the global metal flows to the international resource panel.* UNEP—United Nations Environment Programme.
6. Rüßmann, M., et al. (2015). *Industry 4.0—The future of productivity and growth in manufacturing industries.* https://doi.org/10.1007/s12599-014-0334-4.
7. Lee, J., Bagheri, B., & Kao, H.-A. (2015). A cyber-physical systems architecture for industry 4.0-based manufacturing systems. *Manufacturing Letters, 3,* 18–23. https://doi.org/10.1109/ISORC.2008.25.
8. Nasiri, M., Tura, N., & Ojanen, V. (2017). Developing disruptive innovations for sustainability: A review on Impact of Internet of Things (IOT). In *PICMET '17 - Portland International Conference on Management of Engineering and Technology* (pp. 1–10). https://doi.org/10.23919/PICMET.2017.8125369.

9. Soroka, A., Liu, Y., Han, L., & Haleem, M. S. (2017). Big data driven customer insights for SMEs in redistributed manufacturing. *Procedia CIRP, 63*, 692–697. https://doi.org/10.1016/j.procir.2017.03.319.

10. Dutta, D., Prinz, F. B., Rosen, D., & Weiss, L. (2001). Layered manufacturing: Current status and future trends. *Journal of Computing and Information Science in Engineering, 1*, 60–71. https://doi.org/10.1115/1.1355029.

11. Roy, R. (2000). Sustainable product-service systems. *Futures, 32*, 289–299. https://doi.org/10.1007/978-3-319-70223-0_3.

12. Goedkoop, M. J., van Halen, C. J. G., te Riele, H. R. M., & Rommens, P. J. M. (1999). *Product service systems, ecological and economic basics.*

13. Manzini E., & Vezzoli, C. (2003). A strategic design approach to develop sustainable product service systems: Examples taken from the 'environmentally friendly innovation' Italian prize. *Journal of Cleaner Production, 11*, 851–857. https://doi.org/10.1016/S0959-6526(02)00153-1.

14. Tukker, A., & Tischner, U. (2006). Product-services as a research field: Past, present and future. Reflections from a decade of research. *Journal of Cleaner Production, 14*(17), 1552–1556. https://doi.org/10.1016/j.jclepro.2006.01.022.

15. Sakao, T., & Shimomura, Y. (2007). Service engineering: A novel engineering discipline for producers to increase value combining service and product. *Journal of Cleaner Production, 15*(6), 590–604. https://doi.org/10.1016/j.jclepro.2006.05.015.

16. Pérez-Martínez, S., Giro-Paloma, J., Maldonado-Alameda, A., Formosa, J., Queralt, I., & Chimenos, J. M. (2019). Characterisation and partition of valuable metals from WEEE in weathered municipal solid waste incineration bottom ash, with a view to recovering. *Journal of Cleaner Production, 218*, 61–68. https://doi.org/10.1016/j.jclepro.2019.01.313.

17. Garetti, M., Rosa, P., & Terzi, S. (2012). Life cycle simulation for the design of product-service systems. *Computers in Industry, 63*, 361–369. https://doi.org/10.1016/j.compind.2012.02.007.

18. Baines, T. S., et al. (2007). State-of-the-art in product-service systems. *Proceedings of the Institution of Mechanical Engineers, Part B: Journal of Engineering Manufacture, 221*(10), 1543–1552. https://doi.org/10.1243/09544054JEM858.

19. Fisher, O., Watson, N., Porcu, L., Bacon, D., Rigley, M., & Gomes, R. L. (2018). Cloud manufacturing as a sustainable process manufacturing route. *Journal of Manufacturing Systems, 47*, 53–68. https://doi.org/10.1016/j.jmsy.2018.03.005.

20. Ericson, A., Müller, P., Larsson, T. C., & Stark, R. (2009). Product-service systems–from customer needs to requirements in early development phases. In IPS2—1st CIRP Industrial Product-Service Systems Conference (pp. 1–6), [Online]. Available: http://dspace.lib.cranfield.ac.uk/handle/1826/3612.

Open Access This chapter is licensed under the terms of the Creative Commons Attribution 4.0 International License (http://creativecommons.org/licenses/by/4.0/), which permits use, sharing, adaptation, distribution and reproduction in any medium or format, as long as you give appropriate credit to the original author(s) and the source, provide a link to the Creative Commons license and indicate if changes were made.

The images or other third party material in this chapter are included in the chapter's Creative Commons license, unless indicated otherwise in a credit line to the material. If material is not included in the chapter's Creative Commons license and your intended use is not permitted by statutory regulation or exceeds the permitted use, you will need to obtain permission directly from the copyright holder.

Chapter 2
Circular Business Models Identification

Paolo Rosa, Claudio Sassanelli, and Sergio Terzi

Abstract The main objective of FENIX is demonstrating the benefits coming from the adoption of CE practices through a set of circular business models adequately configured within the project. These CBMs have been selected basing on the three use cases requirements pertaining to different industrial streams (metal powders, 3D-printed jewels and advanced filaments for 3D printing applications). The chapter starts with a literature assessment of both current CBMs and current CBM classification methods. Subsequently, existing CBMs have been mapped basing on the most common classification method (i.e. the ReSOLVE framework), evidencing the most suitable CBMs to be adopted in FENIX. In parallel, a literature assessment of industrial benefits expected from the adoption of CE practices have been implemented. Subsequently, FENIX industrial partners have been interviewed in order to select the most relevant benefits expected from the project. A final comparison of available CBMs and expected benefits allowed to select the most suitable CBMs to be demonstrated in FENIX.

2.1 Current State of the Art on CBMs and Their Classification Methods

Circular Business Models (CBMs) can be considered as the interpretation of circular economy principles within the company's boundaries. Depending on the experts, CBMs (also named as Circular Economy Business Models—CEBMs) can be classified under the wider umbrella of either Green Business Models (GBMs) and/or Sustainable Business Models (SBMs). About this topic, a systematic literature review has been carried out by Rosa et al. [22]. Results unveil that in terms of CBMs the most discussed research areas are (i) practical implementation of CBMs, (ii) challenges related with the adoption of CBMs and (iii) decision-support tools. Considering just

P. Rosa (✉) · C. Sassanelli · S. Terzi
Department of Management, Economics and Industrial Engineering, Politecnico di Milano, Piazza L. da Vinci 32, 20133 Milan, Italy
e-mail: paolo1.rosa@polimi.it

© The Author(s) 2021
P. Rosa and S. Terzi (eds.), *New Business Models for the Reuse of Secondary Resources from WEEEs*, PoliMI SpringerBriefs,
https://doi.org/10.1007/978-3-030-74886-9_2

works on CBM classification methods, it is possible to distinguish three research streams: (i) papers referring to the ReSOLVE framework [24], (ii) papers referring to the Business Model Canvas (BMC) methodology [19] and (iii) papers proposing hybrid models mixing both the previous methods. The ReSOLVE framework [24] aims at supporting companies and governments during the definition of circular economy policies. It identifies six different ways to be circular (e.g. Regenerate, Share, Optimize, Loop, Virtualize and Exchange). Each of them is subsequently detailed in specific actions. Even if the ReSOLVE framework cannot be considered a real classification method, many experts started from it to develop their own models.

Considering the BMC-based classification methods, papers pertaining to this category try to modify the original BMC in order to map circularities.

Considering hybrid models, the experts try to mix the previous classification methods in order to reinforce them. Given the popularity of ReSOLVE and BMC methods, the FENIX project considered them as reference CBM classification methods. Specifically, the ReSOLVE framework has been exploited for the identification of CBM archetypes at macro level. Subsequently, the BMC method has been considered for the detailed description of CBMs at micro level. In addition, a meso classification of CBMs archetypes was adapted from the last OECD's report on CBMs (consisting of fourteen classes considering the full amount of different business models related with circular economy existing in literature) [18].

Considering tables reported by Rosa et al. [22], it is possible to see that some types of CBMs are more frequent than others. The most common CBMs described in literature are represented by recycling practices and use-oriented PSSs. They are followed by bio-based/secondary materials exploitation, reuse and refurbishing/remanufacturing practices, result-oriented and product-oriented PSSs and industrial symbiosis. Not so commonly described in literature are those CBMs related on renewable energies, co-ownership and co-access, repair practices, product dematerialization and new technologies. However, it is evident from the assessed literature the presence of a big research gap in terms of (i) how practically transform linear BMs in circular ones and (ii) how to involve common people in current industrial CBMs. The FENIX project wants to fill in these gaps by proposing practical ways of enabling circular practices in all companies.

2.2 Current State of the Art on Industrial Benefits Related with CBMs

Basing on another systematic literature analysis, Rosa et al. [21] detected and categorized expected benefits related with the adoption of CE. These benefits have been initially classified basing on the triple bottom line of sustainability (i.e. economic, environmental and social) and subsequently grouped in macro categories to ease their detection at industrial level:

- Economic benefits:

    1. Reducing overall costs,
    2. Reducing business risks,
    3. Opening new revenue streams,
    4. Reducing product/process complexity,
    5. Improving competitive advantage,

- Environmental benefits:

    6. Complying with environmental regulations,
    7. Reducing environmental impacts,
    8. Improving resource efficiency,
    9. Improving Supply Chain sustainability,
    10. Reducing Supply Chain,

- Social benefits:

    11. Enhancing reputation and brand value,
    12. Reaching new markets and countries,
    13. Improving health and safety in workplace,
    14. Developing innovative skills and knowledge.

Considering the tables reported by Rosa et al. [21], it is possible to see that some industrial benefits are more frequently considered than others. The most common industrial benefits related with CBMs are resource efficiency, costs and environmental impacts. They are followed by brand reputation, revenue streams, product/process complexity, competitive advantage and supply chain. Not so commonly described in literature are industrial benefits related with business risks, skills and knowledge, new markets, regulations and health and safety. However, it is evident the limited importance given by experts about either social aspect related with CE adoption and the involvement of final users in CBMs. This last point represents one of the key elements for the final selection of the FENIX CBMs.

2.3   Identification of the FENIX Industrial Benefits

In order to select among those detected in the literature review the industrial benefits expected by FENIX partners from the adoption of CBMs, a set of both face-to-face interviews and periodic consultations via phone/web calls have been implemented. The interviews were not based on a pre-defined questionnaire but exploited a set of open questions about both the current and future perspective of some of the FENIX partners. Considering tables reported in Rosa et al. [21], an interesting result is the high importance reached by social aspects (e.g. development of innovative skills and knowledge and enhancement of brand reputation and value) compared with economic (e.g. overall costs reduction) and environmental (e.g. resource efficiency improvement) ones. They are followed by the reduction of the environmental impacts, reduction of business risks, improvement of competitive advantage and supply chain sustainability and provisioning share the same ranking. Subsequently, opening new revenue streams and reducing supply chain complexity seem to be less important. Finally, complying with environmental regulations, reaching new markets and countries, reducing product/process complexity and improving health and safety of workplaces seems to be out of scope for the FENIX partners.

2.4 Identification of the FENIX CBMs

The final decision (based on majority judgement) was to focus on three different CBMs: (1) recycling, (2) result-oriented PSSs and (3) use-oriented PSSs.

- Case 1—The FENIX manufacturing company (recycling-based CBM). This company could produce either a full pilot plant or a specific product. The full pilot plant will either disassemble products, recover materials or manufacture 3D printed components/products. Instead (given a specific AM process), the product could be either a 3D printed jewel, a metal powder for AM processes or an innovative 3D printing filament.
- Case 2—The FENIX service company (use-oriented PSS-based CBM). All the three pilot plants constituting FENIX could act either together or independently (like service providers) focused on a specific process phase. This way, POLIMI's I4.0Lab could act as a provider of assembly/disassembly services for complex products. UNIVAQ's lab could act as a provider of material recovery/refining services. Finally, FCIM, I3DU and MBN labs could act as providers of AM services. However, the plants do not shift in ownership. The provider has ownership, and it is also often responsible for maintenance, repair and control. The customer pays a fee for the use of the plant. He could (or not) have unlimited and individual access (leasing or sharing/pooling).
- Case 3—The FENIX Fablab (result-oriented PSS-based CBM). Here, the final aim is sharing the whole process with final users. This way, the full potential offered by FENIX could be exploited also by private customers willing to implement their ideas. Among the FENIX labs, POLIMI's I4.0Lab is currently the only one already able to adopt this kind of CBM. Here, the PSS still has the three pilot plants as a basis, but the user no longer buys the product produced or the use of the plants. Customers only buy the output of the plants according to the level of use. The provider agrees with the client the delivery of a result. The provider is, in principle, completely free as to how to deliver the result.

In addition, given the high presence of both I4.0 and AM technologies, FENIX could also give a practical demonstration about the adoption of "Exchange" CBMs:

- POLIMI's I4.0Lab will constitute both the initial and final stage of the small-scaled circular supply chain represented within FENIX. This lab is a demonstration plant for the automatic assembly of complex products. FENIX will partially reconfigure it for disassembly needs. Here, the adoption of a Fablab-like CBM is expected to be feasible.

- UNIVAQ's chemical Lab will constitute the central stage of the small-scaled circular supply chain represented within FENIX. This lab: (1) will receive disassembled PCBs from POLIMI, (2) will recover materials from PCBs and (3) will send recovered materials to other partners (e.g. I3DU, MBN and FCIM) for AM-related activities. FENIX will partially reconfigure this Lab for the recovery of selected materials with specific features (e.g. particle's shape, dimension, purity level). Given (i) the high specialization of the lab and (ii) the presence of patented processes, not all the selected CBMs will be feasible.
- FCIM, I3DU and MBN's AM-related Labs will constitute either the semi-final or final stage (depending on the final type of product to be made) of the small-scaled circular supply chain represented within FENIX. If the AM product will be a component of a more complex one, it will be sent to POLIMI's I4.0Lab for the final assembly. FENIX will partially reconfigure these labs basing on specific products/components needs. Given (i) the high specialization of labs and (ii) the presence of patented processes, not all the selected CBMs will be feasible.

2.5   Implementation of the FENIX CBM Assessment Matrixes

Once both CBMs and expected industrial benefits were identified, the final stage was the integration of these views in a common matrix. However, before integrating CBMs and industrial benefits, the focus of analysis must be selected, given the multiple perspective of FENIX considering both pilot plants and final products.

Starting with the pilot plant view, three kinds of PSS-based CBMs can be adopted (e.g. product-oriented, use-oriented and result-oriented ones). Firstly, a product-oriented BM could be adopted in Case 1 (see Sect. 2.4 for details). Secondly, a use-oriented BM could be implemented in Case 3. Finally, a result-oriented BM could be adopted in Case 2.

Considering the final product view, just two out of three kinds of PSS-based CBMs can be adopted (e.g. product-oriented and result-oriented ones). Firstly, a product-oriented BM could be adopted in Case 1 (see Sect. 2.4 for details). Finally, a result-oriented BM could be adopted in Case 2. Given the three pilots implemented within FENIX (all of them starting from electronic scraps sent to the plant by either private or industrial customers), six different CBMs could be adopted. Two of them are related with the production of green metal powders for AM processes, two are related with the production of 3D printed jewels from green precious metals and two are related with the production of either Additive Manufacturing (AM) materials or 3D printing filaments from wasted materials. What is evident from tables reported in Rosa et al. [21] is the absence of a CBM offering better chances to fill in great part of the expected industrial benefits. Instead, use-oriented and result-oriented PSSs will allow to better cope with social aspects related with CE.

2.6   Conclusions

This chapter presented the three Circular Business Models (CBMs) to be adopted within the FENIX project. These CBMs were identified in product-oriented, use-oriented and result-oriented PSSs. For their identification, a multi-perspective procedure has been adopted. First, a state-of-the-art analysis allowed to define the most common types of CBMs and their classification methods. Secondly, a set of dedicated interviews with all the FENIX partners allowed the definition of the most important industrial benefits expected from the adoption of circular practices. Together, the integration of both the scientific and industrial perspective allowed the identification of the most suitable CBMs to consider within the FENIX project, distinguishing among CBMs related to the pilot plant itself and CBMs related with specific products coming from the pilot plant.

References

1. Antikainen, M., & Valkokari, K. (2016). A framework for sustainable circular business model innovation. *Technology Innovation Management Review, 5*(7), 1–8. https://doi.org/10.22215/timreview/1000.
2. Bocken, N. M. P., de Pauw, I., Bakker, C. A., & van der Grinten, B. (2016). Product design and business model strategies for a circular economy. *Journal of Industrial and Production Engineering, 33*(5), 308–320. https://doi.org/10.1080/21681015.2016.1172124.
3. Bocken, N. M. P., Short, S. W., Rana, P., & Evans, S. (2014). A literature and practice review to develop sustainable business model archetypes. *Journal of Cleaner Production, 65*, 42–56. https://doi.org/10.1016/j.jclepro.2013.11.039.
4. Charter, M. (2016). Circular economy business models. *Sustainable Innovation 2016*, 64–69.
5. Chiappetta Jabbour, C. J., de Sousa, L., Jabbour, A. B., Sarkis, J., & Godinho Filho, M. (2019). Unlocking the circular economy through new business models based on large-scale data: An integrative framework and research agenda. *Technological Forecasting & Social Change, 144*, 546–552. https://doi.org/10.1016/j.techfore.2017.09.010.
6. Chiaroni, D., Urbinati, A., & Chiesa, V. (2016). Circular economy business models: Towards a new taxonomy of the degree of circularity. In *The XXVII Edition of the Annual Scientific Meeting of the Italian Association of Management Engineering (AiIG), Higher Education and Socio-Economic Development* (pp. 1–27). https://doi.org/10.1016/j.jclepro.2017.09.047.
7. Haanstra, W., Toxopeus, M. E., & van Gerrevink, M. R. (2017). Product life cycle planning for sustainable manufacturing: Translating theory into business opportunities. *Procedia CIRP, 61*, 46–51. https://doi.org/10.1016/j.procir.2016.12.005.
8. Heyes, G., Sharmina, M., Mendoza, J. M. F., Gallego-Schmid, A., & Azapagic, A. (2018). Developing and implementing circular economy business models in service-oriented technology companies. *Journal of Cleaner Production, 177*, 621–632. https://doi.org/10.1016/j.jclepro.2017.12.168.
9. Janssen, K. L., & Stel, F. (2017). Orchestrating partnerships in a circular economy—A working method for SMEs. In *The XXVIII ISPIM Innovation Conference* (pp. 1–17).
10. Lewandowski, M. (2016). Designing the business models for circular economy—Towards the conceptual framework. *Sustainability, 8*(43), 1–28. https://doi.org/10.3390/su8010043.
11. Lozano, R., Witjes, S., van Geet, C., & Willems, M. (2016). *Collaboration for circular economy: Linking sustainable public procurement and business models.* https://doi.org/10.13140/RG.2.2.36081.68969.

12. Manninen, K., Koskela, S., Antikainen, R., Bocken, N. M. P., Dahlbo, H., & Aminoff, A. (2018). Do circular economy business models capture intended environmental value propositions? *Journal of Cleaner Production, 171*, 413–422. https://doi.org/10.1016/j.jclepro.2017.10.003.
13. Mendoza, J. M. F., Sharmina, M., Gallego-Schmid, A., Heyes, G., & Azapagic, A. (2017). Integrating backcasting and eco-design for the circular economy: The BECE framework. *Journal of Industrial Ecology, 21*(3), 526–544. https://doi.org/10.1111/jiec.12590.
14. Morioka, S. N., Bolis, I., & Monteiro de Carvalho, M. (2018). From an ideal dream towards reality analysis: Proposing Sustainable Value Exchange Matrix (SVEM) from systematic literature review on sustainable business models and face validation. *Journal of Cleaner Production, 178*, 76–88. https://doi.org/10.1016/j.jclepro.2017.12.078.
15. Nerurkar, O. (2017). A framework of sustainable business models. *Indian Journal of Economics and Development, 5*(1), 1–6.
16. Nußholz, J. L. K. (2017). Circular business models: Defining a concept and framing an emerging research field. *Sustainability, 9*(1810), 1–16. https://doi.org/10.3390/su9101810.
17. Nußholz, J. L. K. (2017). Circular business model framework : Mapping value creation architectures along the product lifecycle. In *PLATE 2017—Product Lifetimes And The Environment Conference* (pp. 1–8).
18. OECD—European cooperation and economic development organization. (2017). *New business models for the circular economy—Opportunities and challenges from a policy perspective* (Issue June).
19. Osterwalder, A., & Pigneur, Y. (2010). *Business model generation*. Wiley. https://doi.org/10.1017/CBO9781107415324.004.
20. Prendeville, S., & Bocken, N. M. P. (2016). Sustainable business models through service design. In *GCSM 2016—14th Global Conference on Sustainable Manufacturing* (pp. 292–299). https://doi.org/10.1016/j.promfg.2017.02.037.
21. Rosa, P., Sassanelli, C., & Terzi, S. (2019). Circular business models versus circular benefits: An assessment in the waste from electrical and electronic equipments sector. *Journal of Cleaner Production, 231*, 940–952. https://doi.org/10.1016/j.jclepro.2019.05.310.
22. Rosa, P., Sassanelli, C., & Terzi, S. (2019). Towards circular business models: A systematic literature review on classification frameworks and archetypes. *Journal of Cleaner Production, 236*(117696), 1–17. https://doi.org/10.1016/j.jclepro.2019.117696.
23. Stratan, D. (2017). Success factors of sustainable social enterprises through circular economy perspective. *Visegrad Journal on Bioeconomy and Sustainable Development, 6*(1), 17–23. https://doi.org/10.1515/vjbsd-2017-0003.
24. The Ellen MacArthur Foundation. (2015). *Towards a circular economy: Business rationale for an accelerated transition*, April 3, 2012.
25. Tolio, T., Bernard, A., Colledani, M., Kara, S., Seliger, G., Duflou, J. R., Battaia, O., & Takata, S. (2017). Design, management and control of demanufacturing and remanufacturing systems. *CIRP Annals—Manufacturing Technology, 66*(2), 585–609. https://doi.org/10.1016/j.cirp.2017.05.001.
26. Urbinati, A., Chiaroni, D., & Chiesa, V. (2017). Towards a new taxonomy of circular economy business models. *Journal of Cleaner Production, 168*, 487–498. https://doi.org/10.1016/j.jclepro.2017.09.047.
27. Witjes, S., & Lozano, R. (2016). Towards a more Circular economy: Proposing a framework linking sustainable public procurement and sustainable business models. *Resources, Conservation and Recycling, 112*, 37–44. https://doi.org/10.1016/j.resconrec.2016.04.015.

Open Access This chapter is licensed under the terms of the Creative Commons Attribution 4.0 International License (http://creativecommons.org/licenses/by/4.0/), which permits use, sharing, adaptation, distribution and reproduction in any medium or format, as long as you give appropriate credit to the original author(s) and the source, provide a link to the Creative Commons license and indicate if changes were made.

The images or other third party material in this chapter are included in the chapter's Creative Commons license, unless indicated otherwise in a credit line to the material. If material is not included in the chapter's Creative Commons license and your intended use is not permitted by statutory regulation or exceeds the permitted use, you will need to obtain permission directly from the copyright holder.

Chapter 3
Circular Economy Performance Assessment

Roberto Rocca, Claudio Sassanelli, Paolo Rosa, and Sergio Terzi

Abstract The main aim of the FENIX project is the development of new business models and industrial strategies for three novel supply chains in order to enable value-added product-services. Through a set of success stories coming from the application of circular economy principles in different industrial sectors, FENIX wants to demonstrate in practice the real benefits coming from its adoption. In addition, Key Enabling Technologies (KETs) will be integrated within the selected processes to improve the efficient recovery of secondary resources. This chapter focuses on the definition of a novel Circular Economy Performance Assessment (CEPA) methodology to be adopted within the FENIX project. This implementation activity has been done into two steps. From one side, a state-of-the-art analysis of existing CE methodologies and related KPIs has been executed and the most common circularity assessment methods (and KPIs) have been identified. Subsequently, a totally new CEPA methodology has been developed starting from the findings coming from the literature. This methodology, together with classic LCA and LCC methods, will be exploited for the quantitative assessment of CBMs.

3.1 State of the Art on Circular Economy Performance Assessment Methods

Circular Economy (CE) research is continuously evolving. Especially in the last years, this led both researchers and practitioners to understand how to measure and quantify its impacts in a real context. Trying to summarize the findings coming from an extensive literature review reported in [4], there is a strong orientation of CEPA methodologies on the environmental aspect of the Triple Bottom Line (TBL) of sustainability. Indeed, all the contributions involve the environmental perspective, either alone (37.7%) or combined with the economic one (17%) or embedded in the

R. Rocca (✉) · C. Sassanelli · P. Rosa · S. Terzi
Department of Management, Economics and Industrial Engineering, Politecnico di Milano,
Piazza L. da Vinci 32, 20133 Milan, Italy
e-mail: roberto.rocca@polimi.it

© The Author(s) 2021
P. Rosa and S. Terzi (eds.), *New Business Models for the Reuse of Secondary Resources from WEEEs*, PoliMI SpringerBriefs,
https://doi.org/10.1007/978-3-030-74886-9_3

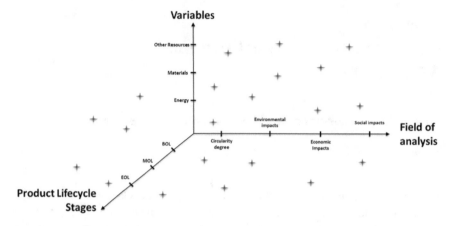

Fig. 3.1 Positioning framework

entire triple perspective (45.3%). This environmental perspective was also differentiated among energy, materials and pollution, or a combination of them. Once again, the focus was on one element (i.e. materials and pollution). This confirms the importance of such variables in the circularity performance context, since a continuous flow of technical and biological materials through the 'value circle' is considered in CE [5].

Starting from these categorizations, POLIMI's team created a framework to position existing methodologies in order to map the existing gaps in literature (see Fig. 3.1). The framework is constituted by three axes: (i) Product Lifecycle Stages, for mapping lifecycle phases considered; (ii) Variables, for mapping the type of variables considered and measured; and (iii) Field of analysis, for mapping the perspective used to analyse variables.

What emerges from the mapping process is the relatively low amount of industries considering benefits related with biological models and resource flows as a source of inspiration for new measures able to reduce environmental impacts and, in parallel, generate economic savings. To this aim, it is necessary to work both at system and single product level (i.e. company level or entire product lifecycle level) at the same time, going into detail at single production phase and single resource flow. For this reason, a quantitative (product-oriented) assessment model has been proposed to calculate the circularity performance.

3.2 The Circular Economy Performance Assessment Methodology

The Circular Economy Performance Assessment (CEPA) methodology proposed by POLIMI is composed by three different sub-methodologies. Each of them is related to three different fields of analysis: (i) a Circularity Product Assessment (CPA), (ii) a Circularity Cost Assessment (CCA) and (iii) a Circularity Environmental Assessment (CEA). The first sub-methodology is presented in this chapter, while the other ones are just mentioned in a qualitative manner.

Firstly, CPA allows to both calculate the circular share of resource flows used during the product life cycle and obtain an exhaustive index (KPI) about the circular percentage share of the product compared to total resources used (Circularity Product Indicator, CPI). Given its dependency on the resources type exploited for the creation of a generic product, this methodology can be exploited to compare different scenarios and assess the most virtuous ones in terms of resource flows maximization. Secondly, CCA allows to analyse and quantify the economic benefits related to CE, always referring to a well-defined product. It can be exploited to both calculate the cost savings generated by the triggering of materials and other resources circularity and evaluate the economic savings related to energy circularity. Finally, CEA allows to evaluate the environmental benefits resulting from the use of a CBM. Here the focus is quantifying emissions and other forms of pollution avoided by triggering the resources flows circularity present throughout the entire life cycle. This methodology consists in the association of a "weight" to all the environmental impacts characterizing each circular resource flow, in order to be able to calculate the difference with the environmental impacts of the corresponding linear system.

Outputs of the CEPA methodology consist in a set of specific KPIs regarding resources circularity degree present within the product life cycle and the quantification of economic and environmental benefits related with CE. They can be used in different application fields:

- Creation of certification standards related to resource flows circularity
- Decision-support of new products (e.g. defining Design for CE guidelines)
- Circular scenarios comparison (e.g. based on circularity levels and benefits) of both new and existing products
- Internal reporting and benchmarking.

In general terms, the main principle on which the CEPA methodology developed by POLIMI is based on is the Material Flow Analysis (MFA). MFA is a systematic assessment of stocks and flows of materials within a system defined in space and time. Because of the physical law of conservation of matter, the results of an MFA can be controlled by a simple material balance comparing inputs, stocks and outputs of a process. This characteristic of MFA makes the method attractive as a decision-support tool in resource management, waste management and environmental management [1, 2].

3.3 Circularity Product Assessment (CPA) Methodology

The objective of CPA is quantifying the circularity level of each resource involved within a product lifecycle. Given 100% as the input quantity of a given resource "k" in a generic product lifecycle phase "p", X% of this input will end up in the output of that activity, Y% will be discarded and—in case of circularities—Z% will be reused either within the same system or in a different one. Therefore, the generic constraint to be considered is:

$$X\% + Y\% + Z\% = 100\% \text{ of resource } k \text{ in phase } p.$$

This calculation process is carried out for all the resources and phases of the analysed system, trying to limit the analysis to the product lifecycle. Unlike the objectives of environmental impact assessment models (e.g. LCA), CPA aims to identify and quantify circularities present in a system. With the term "circularity" we refer to those feedback resource flows (of any kind) falling retroactively in the system (the same or another) (Fig. 3.2).

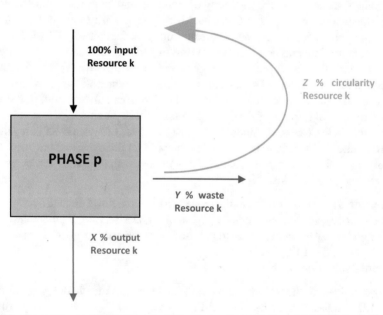

Fig. 3.2 Flow schematization

3.3.1 CPA Phase 1—Objectives Definition and Settings

CPA methodology is composed by four different phases. Phase 1 identifies the context of analysis. It is very important to understand from the beginning what kind of study must be carried out, as it is associated with both different modelling principles and methodological choices. Regarding the modelling principles to be used, we refer to LCA guidelines [6, 7], given the similarity between phase 1 of the proposed model and that of LCA. The modelling principles are two: (i) the "attributional" modelling, describing the system in a static way; and (ii) the "consequential" modelling, describing the system as a result of the analysed decisions, then inserting it into a dynamic sphere.

Furthermore, the context of analysis may vary depending on whether the analysis is undertaken to support (or not) any decision and basing on the types of process changes in the analysed system. Considering LCA guidelines, Phase 1 defines:

- System's boundaries (the part of lifecycle to be considered)
- Functional units and reference flows (units of measure needed to quantify performances and quantity of product necessary to satisfy the chosen functional unit, respectively)
- Data characteristics (e.g. precision, completeness, representativeness, etc.)
- Allocation procedures and multi-process case resolution (e.g. closed-loop versus open-loop allocations)
- Hypotheses and limitations.

3.3.2 CPA Phase 2—Inventory Analysis and Resource Flows Decomposition

Phase 2 includes the compilation and quantification of inputs and outputs of each phase for a product/functional unit during its lifecycle. Data must be collected for each lifecycle phase included within the system's boundaries and referred to each functional unit. Coherently with the identified system's boundaries, the phases analysed will be p (ranging from 1 to P).

If the product, or a part of it, is repaired or remanufactured (because it is convenient at the economic level and for the resources used), we can talk about maintenance circularity. According to the cardinal principles of CE, the repair of a product triggers a retroactive flow in the technical sphere of the system. This allows a saving of resources compared to the case in which a new product was created and in addition allows the extension of the product lifecycle. For this reason, the circularity level is quantified as the ratio between saved resources and those used in case of a new production. Since not all products can be repaired (e.g. fast-moving consumer goods) and repairable ones can have an uncertainty of non-repairability, the maintenance phases present a binary variable that is activated only if needed. Then, another uncertainty to consider is related to the lengthening of the product lifecycle. To this aim, a coefficient expresses the extension of the lifecycle due to maintenance, compared to the average lifespan of the product itself.

- $f [0, 1, ..., F]$ = number of repairs
- A = binary variable whose value is:

  - 1 if $f > 0$ (at least one repair for the product)
  - 0 if $f = 0$ (no repair for the product)

- $z_f = \frac{(f + 1)\text{-th useful life}}{\text{average useful life}}$ = coefficient expressing the product lifecycle extension thanks to maintenance and repairs. f is the number of times the product is repaired. Consequently, "(f + 1)-th useful life" represents the useful lifespan after the f-th maintenance, so "[(f + 1)-th useful life]/(average useful life))]" indicates the product lifecycle extension after the f-th maintenance (with respect to its average duration).

After having described all the system phases, the methodology moves on to the inventory phase of all the resources used to create the product: (i) energy flows (electricity and thermal energy); (ii) material flows (materials constituting the product); (iii) complementary resource flows (e.g. water, cooling fluids, chemical additives, consumables, etc.).

These three types of flows will be quantified and allocated in the respective phases, so allowing the subsequent calculation of the different types of circularity. The definition of precise system's boundaries does not necessarily imply the study of a closed system. In fact, among the different types of material flows and resources considered there are also those coming from other systems. So, also the possible interactions of lifecycles of other products must be considered. This logic creates a bridge between the concept of product lifecycle and Industrial Symbiosis (IS). In particular, the following types of circularity will be analysed:

- Electric or thermal flows from renewable energy sources
- Thermal energy flows from thermal recovery
- Electric or thermal flows from recovery of discarded materials and resources
- Material flows (or other non-virgin resources) in input from other systems
- Material flows (or other non-virgin resources) in input from the same system (short-range if coming from the same phase p, long-range if coming from another phase p, or from the end-of-life phase)
- Material flows (or other non-virgin output resources) intended for re-use in the same system or in other systems
- Resource flows saved as a result of maintenance and repair activities.

3.3.3 CPA Phase 3—Weights and Indexes Calculation

Phase 3 calculates weights and indexes used in CPA. In particular, the focus is on "physical" weight of materials (and other resources) and on weight of each phase in terms of resources used (energy, materials and other resources). Subsequently, the recyclability characteristics of materials have been considered to calculate their potential reuse. The following weights and indices are proposed:

- W_p^E: energy weight in phase p
- $W_{m,p}^{M,P}$: m-th material weight in phase p
- $W_{r,p}^{R,P}$: r-th resource weight in phase p
- W_m^M: m-th material relative weight
- W_r^R: r-th resource relative weight
- CRI_m: composed recyclability index of the m-th material

W_p^E: Weight of energy consumed in phase p out of the total energy of the system

$$W_p^E = \frac{EE_p + TE_p}{E_{system}}$$

In case of maintenance phase ($A = 1$; $p = f1 \rightarrow ff$), the energy weight of the generic maintenance phase f will become:

$$W_{f_f}^E = \frac{E\_maint_f}{E\_system}$$

- EE_p = Kwh of Electricity consumed in phase p
- TE_p = Kwh of Thermal Energy consumed in phase p
- $EE = \Sigma_{p=1}^P (EE_p)$ = Kwh of Electricity consumed in the lifecycle, including consumption of all recycling activities necessary to recover resources
- $TE = \Sigma_{p=1}^P (TE_p)$ = Kwh of Thermal Energy consumed in the lifecycle, including the consumption of all recycling activities necessary to recover resources.

Total energy within the lifecycle:

$$E\_system = EE + TE + \left(A * \sum_{f=1}^F E\_maint_f \right)$$

W_p^E represents the "energy weight of phase p". This is the amount of energy consumed in phase p out of the total energy consumed within system's boundaries. The total energy balance (denominator of the formula) is obtained by adding together any type of energy consumption present in the product lifecycle. It refers to its functional unit. In addition to energy consumptions related with production, energy consumptions related to any resource recovery activity are also considered. The amount of energy consumed is subtracted from the amount of energy generated if the product (or part of it) is destined to energy recovery, thus creating an energetic circularity.

$W_{m,p}^{M,P}$: Weight of the m-th input material in phase p out of the total m-th material used in the system

$$W_{m,p}^{M,P} = \frac{MF\_in_{m,p}}{MF\_in_m + \left(A * \sum_{f=1}^{F} MF\_maint\_in_{m,f}\right)}$$

In case of maintenance phase ($A = 1$; $p = f1 \rightarrow ff$), the material weight of the generic maintenance phase f will become:

$$W_{m,f_f}^{M,F} = \frac{MF\_maint\_in_{m,f}}{MF\_in_m + \left(A * \sum_{f=1}^{F} MF\_maint\_in_{m,f}\right)}$$

$W_{r,p}^{R,P}$: Weight of the r-th resource in input in phase p out of the total resources used in the system

$$W_{r,p}^{R,P} = \frac{RF\_in_{r,p}}{RF\_in_r + \left(A * \sum_{f=1}^{F} RF\_maint\_in_{r,f}\right)}$$

In case of maintenance phase ($A = 1$; $p = f1 \rightarrow ff$), the resource weight of the generic maintenance phase f will become:

$$W_{r,f_f}^{R,F} = \frac{RF\_maint\_in_{r,f}}{RF\_in_r + \left(A * \sum_{f=1}^{F} RF\_maint\_in_{r,f}\right)}$$

W_m^M: Weight of materials used in input out of the total input materials used to make the product

$$W_m^M = \frac{MF\_in_m + \left(A * \sum_{f=1}^{F} MF\_maint\_in_{m,f}\right)}{\sum_{m=1}^{M}(MF\_in_m) + \left(A * \sum_{f=1}^{F} \sum_{m=1}^{M} MF\_maint\_in_{m,f}\right)}$$

MF_in_m: Mass of m-th material in input within the production process.
[m: from 1 to M → materials].

W_m^M represents the "weight of the m-th material" out of the total materials embedded in the product. This is the first weight used for the material's circularity calculation. It represents a "physical" index since it is a ratio between masses.

W_r^R: Weight of input resources used out of the total input resources

$$W_r^R = \frac{RF\_in_r + \left(A * \sum_{f=1}^{F} RF\_maint\_in_{r,f}\right)}{\sum_{r=1}^{R}(RF\_in_r) + \left(A * \sum_{f=1}^{F} \sum_{m=1}^{M} RF\_maint\_in_{r,f}\right)}$$

RF_in_r: Mass (or volume) of r-th resource in input within the production process.
[r: from 1 to R → other resources].

W^R_r represents the "weight of the r-th resource" out of the total amount of other resources used to make the product. Constraints dictated by the mass and energy balances impose that $\sum^P_{p=1}$ of both W^E_p, $W^{M,P}_{m,p}$ and $W^{R,P}_{r,p}$ will be equal to 1. For the same reason, also $\sum^M_{m=1}$ of W^M_m and $\sum^R_{r=1}$ of W^R_r will be equal to 1.

In order to calculate material's recyclability, both physical properties (and their evolution) and economic value of materials must be considered. From one side, the Substitution Rate (R_s) is a good approximation of the physical property evolution of recycled materials compared to the number of reuses [3]. If not available from standard lists, it can be easily calculated for a generic m-th material. From another side, the economic value of materials can be estimated through the Recyclability Index [8]. Here, the quality of the recycled material can offer a good estimate of the final material's value. The higher the quality, the lower the difference between virgin and secondary materials market values. In order to calculate material's recyclability (R), the following values must be considered:

- V_m: minimum material value (€/kg) before its processing or forming.
- V_r: material's residual value (€/kg) after its primary use and before its recycling.
- V_p: material's post-recycling value (€/kg) before its processing or forming. If a material has both high recyclability and high-quality, V_p will be close to V_m.

Therefore, the Recyclability Index (R) is:

$$R = \frac{V_p}{V_m}$$

If $V_p \ll V_m$, the material is reused, sent to landfill or used for energy recovery. If $V_p \ll V_m$, the recycling process is not economically sustainable.

Starting from the Recycling Index, a binary variable (β) is added to the calculation in order to check the economic sustainability of a recycled material. $\beta = 1$ means that it is economically feasible to use recycled material. Otherwise, $\beta = 0$. However, the economic aspects associated to resource circularity will be considered by the CCA methodology. Once calculated the substitution rate (R_s) and the Recyclability Index (R), it is possible to calculate the Composed Recyclability Index (CRI):

$$CRI = \beta * R_s$$

Even if one between β and R_s is null, CRI is null because for either technical or economic reasons it is not possible (or profitable) to use the recycled material. Although materials have a specific weight, CRI does not consider the danger and toxicity degrees of materials and other resources present in the production process. This calculation will be done by the CEA methodology.

3.3.4   CPA Phase 4—Circularity Indexes Calculation

Phase 4 calculates the circularity indexes for each type of resource. Circular shares are weighted in each system phase and grouped under a single index named Circularity Product Indicator (CPI). Then, circularity yields are calculated both for materials and other resources and their values come from the relationship between "generated" and "absorbed" circularities.

EEC_p: Percentage share of electricity from renewable energies used in phase p out of the total energy consumed in phase p

$$EEC_p = \frac{EE\_R_p}{(EE_p + TE_p)}$$

TEC_p: Percentage share of thermal energy from renewable energies or recovered through heat recovery out of the total energy consumed in phase p

$$TEC_p = \frac{TE\_R_p}{(EE_p + TE_p)}$$

ECI_p: Energy Circularity Indicator of the product in phase p

$$ECI_p = W_p^E * (EEC_p + TEC_p) * 100$$

EC_maint_f = Energy Maintenance Circularity of the f-th maintenance

$$EC\_maint_f = \frac{E\_saved_f}{E\_system} = \left\{ 1 - \left[\frac{E\_maint_f}{E\_system} \right] \right\} * 100$$

E_saved_f = Total kWh saved doing the f-th maintenance, compared to a new product cycle.

$$E\_saved_f = (E\_system) - E\_maint_f$$

ECI: Overall Energy Circularity Indicator

$$ECI = \sum_{p=1}^{P} (ECI_p) + A * \sum_{f=1}^{F} (EC\_maint_f * W_{f_f}^E * z_f)$$

$MCI_{m,p}$ = Material Circularity Indicator (absorbed circularity) of the m-th material in phase p.

$$MCI_{m,p} = \frac{\left(MFC\_in_{m,p}^{short} + MFC\_in_{m,p}^{long} + MFC\_in_{m,p}^{col} + MFC\_in_{m,p}^{OS} \right)}{MF\_in_{m,p}}$$

MC_maint$_{m,f}$ = Material Maintenance Circularity of the m-th material for the f-th maintenance activity

$$\text{MC\_maint}_{m,f} = \left[1 - \frac{\text{MF\_maint\_in}_{m,f}}{\text{MF\_in}_m}\right]$$

This way, savings coincide with masses of the m-th material not used to repair the product, but to build a new one. MF_maint_in$_{m,f}$ is the mass of m-th material in input in the maintenance phase f. It is the mass used for making the spare part or for direct repair.

MCI$_m$ = Material Circularity Indicator (absorbed circularity) of the m-th material

$$\text{MCI}_m = \sum_{p=1}^{P}\left[\text{MCI}_{m,p} * W_{m,p}^{M,P}\right] + A * \sum_{f=1}^{F}\left[\text{MC\_maint}_{m,f} * W_{m,f_f}^{M,F} * z_f\right]$$

MCI = Overall Material Circularity Indicator (absorbed circularity)

$$\text{MCI} = \sum_{m=1}^{M}\left[\left(\text{MCI}_m * W_m^M * \text{CRI}_m\right) * 100\right]$$

RCI$_{r,p}$ = Resource Circularity Indicator (absorbed circularity) of the r-th resource in phase p

$$\text{RCI}_{r,p} = \frac{\left(\text{RFC\_in}_{r,p}^{short} + \text{RFC\_in}_{r,p}^{long} + \text{RFC\_in}_{r,p}^{eol} + \text{RFC\_in}_{r,p}^{OS}\right)}{\text{RF\_in}_{r,p}}$$

RC_maint$_{r,f}$ = Resource Maintenance Circularity of the r-th resource for the f-th maintenance activity

$$\text{RC\_maint}_{r,f} = \left[1 - \frac{\text{RF\_maint\_in}_{r,f}}{\text{RF\_in}_r}\right]$$

This way, savings coincide with masses (or volume) of the r-th resource not used to repair the product, but to build a new one. RF_maint_in$_{r,f}$ is the mass (or volume) of r-th resource in input in the maintenance phase f.

RCI$_r$ = Resource Circularity Indicator (absorbed circularity) of the r-th resource

$$\text{RCI}_r = \sum_{p=1}^{P}\left[\text{RCI}_{r,p} * W_{r,p}^{R,P}\right] + A * \sum_{f=1}^{F}\left[\text{RC\_maint}_{r,f} * W_{r,f_f}^{R,F} * z_f\right]$$

RCI = Overall Resource Circularity Indicator (absorbed circularity)

$$RCI = \sum_{r=1}^{R}[(RCI_r * W_r^R) * 100]$$

CPI can have values between 0 and 1 and must be zero if ECI, MCI and RCI are simultaneously null (no type of circularity in the system) and one in case the value of ECI, MCI and RCI is at the same time one (totally circular system).

Considering the problem from a geometrical point of view, it is possible to consider ECI, MCI and RCI as three independent variables in a three-dimensional space ("equally important"). We can, therefore, consider the "total circularity of the system" as a sphere centred in the origin of ECI, MCI and RCI axes, with a radius K:

$$K = \sqrt{ECI^2 + MCI^2 + RCI^2}$$

If we consider the maximum radius of the sphere ($K_{max} = \sqrt{3}$), CPI will be:

$$CPI = \frac{K}{\sqrt{3}} = \frac{\sqrt{ECI^2 + MCI^2 + RCI^2}}{\sqrt{3}} * 100$$
$$0 \le CPI \le 1.$$

CPI quantifies the generated circularity (i.e. resources available for the same system or others) less the absorbed one (i.e. received from the same system or from others). In order to make CPI as complete as possible, we need to consider how much a system can generate flows of reusable resources, compared to those received in input. They are measured through a set of circularity performance indexes (e.g. η_{EC} = energy circularity performance, η_{MC} = material circularity yield, η_{RC} = resource circularity yield).

η_{EC} quantifies the generated circular energy flows compared with the absorbed one. Considering that:

- $MFC\_out_{m,p}^{en\_rec}$: Mass of m-th material discarded by phase p and sent to energy recovery
- $RFC\_out_{r,p}^{en\_rec}$: Mass of r-th resource rejected by phase p and send to energy recovery

If ($p = EoL$), then:

$MFC\_out_{m,p}^{en\_rec} = MFC\_out_{m,eol}^{en\_rec} eRFC\_out_{r,p}^{en\_rec} = RFC\_out_{r,eol}^{en\_rec}$: Mass of m-th material (or r-th resource) discarded by EoL phase and sent to energy recovery. This mass is part of the finished product at EoL phase.

If ($p = f_f$), then:

$MFC\_out_{m,p}^{en\_rec} = MFC\_out_{m,ff}^{en\_rec} eRFC\_out_{r,p}^{en\_rec} = RFC\_out_{r,ff}^{en\_rec}$: Mass of m-th material (or r-th resource) discarded by maintenance activity f_f and sent to energy recovery.

LHV_m = Lower heating value of the m-th material sent to energy recovery.

LHV_r = Lower heating value of the r-th resource sent to energy recovery.

$\eta_m^{en\_rec}$ = Energy recovery process yield, where m-th material is used.

$\eta_r^{en\_rec}$ = Energy recovery process yield where r-th resource is used.

E_rec^M = Energy circularity generated by discarded materials sent to energy recovery

$$E\_rec^M = \left\{ \left[\sum_{p=1}^{P} \sum_{m=1}^{M} \left(MFC\_out_{m,p}^{en\_rec} * LHV_m * h_m^{en\_rec} \right) \right] \right.$$
$$\left. + \left[\sum_{m=1}^{M} \left(MFC\_out_{m,eol}^{en\_rec} * LHV_m * h_m^{en\_rec} \right) \right] \right\}$$

$E\_rec\_max^M$ = Maximum energy circularity, potentially generable from discarded materials sent to energy recovery

$$E\_rec\_\overset{M}{max} = \left\{ \left[\sum_{p=1}^{P} \sum_{m=1}^{M} \left(MF\_out_{m,p} * LHV_m \right) \right] + \left[\sum_{m=1}^{M} \left(MF\_FP_{m,eol} * LHV_m \right) \right] \right\}$$

E_rec^R = Energy circularity generated by other resources discarded and sent to energy recovery

$$E\_rec^R = \left\{ \left[\sum_{p=1}^{P} \sum_{r=1}^{R} \left(RFC\_out_{r,p}^{en\_rec} * LHV_r * h_r^{en\_rec} \right) \right] \right.$$
$$\left. + \left[\sum_{r=1}^{R} \left(RFC\_out_{r,eol}^{en\_rec} * LHV_r * h_r^{en\_rec} \right) \right] \right\}$$

$E\_rec\_max^R$ = Maximum energy circularity, potentially generable by other resources discarded and sent to energy recovery

$$E\_rec\_\overset{R}{max} = \left\{ \left[\sum_{p=1}^{P} \sum_{r=1}^{R} \left(RF\_out_{r,p} * LHV_r \right) \right] + \left[\sum_{r=1}^{R} \left(RF\_FP_{r,eol} * LHV_r \right) \right] \right\}$$

ECI_out = Energy Circularity Indicator regarding circularities generated

$$ECI_{out} = \frac{(E\_rec^M + E\_rec^R)}{(E\_rec\_max^M + E\_rec\_max^R)}$$

If $ECI \neq 0$ it is possible to calculate:

η_{EC} = Energy Circularity Yield of the system

$$\eta_{EC} = \frac{ECI\_out}{ECI}$$

η_{MC} considers the generated material circularity performance compared to the absorbed one. For materials going in other systems, after being used for making the product, we need to consider:

- t_m: number of times the m-th material has been used
- T_{m_max}: maximum number of times the m-th material can be used

These measures allow to assign a "temporal weight" to the m-th material destined for recycling (or re-use) in other systems: $W_m^T = (T_{m\_max} - t_m)/T_{m\_max}$.

The higher this ratio is, the greater the number of potential reuse cycles of the m-th material in other systems (or in the same system). For materials flows going to energy recovery or to landfill $W_m^T = 1$.

$MFC\_out_{m,p}^{OS}$: Mass of m-th material discarded by phase p and used in other systems.

$MFC\_out_{m,p}^{SS}$: Mass of m-th material discarded from phase p and reused in the same system (in phase p or other phases).

$MFC\_out_{m,p}^{en\_rec}$: Mass of m-th material discarded from phase p and sent to energy recovery.

$MFC\_out_{m,eol}^{OS}$: Mass of m-th material discarded from the EoL and reused in other systems.

$MFC\_out_{m,eol}^{SS}$: Mass of m-th material discarded from the EoL and reused in the same system (in phase p or in other phases).

$MFC\_out_{m,eol}^{en\_rec}$: Mass of m-th material discarded by the EoL and sent to energy recovery.

$MFC\_out_{m,eol}$: Mass of m-th material discarded from the EoL and reused in the same system or outside it.

$MC\_out_{m,p}^{OS}$ = Generated circularity of the m-th material in phase p for other systems

$$MC\_out_{m,p}^{OS} = \frac{MFC\_out_{m,p}^{OS}}{MF\_out_{m,p}}$$

$MC\_out_{m,p}^{SS}$ = Generated circularity of the m-th material in phase p for the same system

$$MC\_out_{m,p}^{SS} = \frac{MFC\_out_{m,p}^{SS}}{MF\_out_{m,p}}$$

$MC\_out_m^{OS}$ = Generated circularity of the m-th material for other systems

$$MC\_out_m^{OS} = \sum_{p=1}^{P} \left[\frac{MFC_{out\ m,p}^{OS}}{MF\_out_m + MF_{FPm}} \right] + \left[\frac{MFC\_out_{m,eol}^{OS}}{MF\_out_m + MF\_FP_m} \right]$$

$MC\_out_m^{SS}$ = Generated circularity of the m-th material for the same systems

$$MC\_out_m^{SS} = \sum_{p=1}^{P} \left[\frac{MFC_{out\ m,p}^{SS}}{MF\_out_m + MF\_FP_m} \right] + \left[\frac{MFC\_out_{m,eol}^{SS}}{MF\_out_m + MF\_FP_m} \right]$$

MCI_out^{OS} = Material Circularity Indicator regarding the circularity generated by the product for other systems

$$MCI\_out^{OS} = \sum_{m=1}^{M} \left(MC\_out_m^{OS} * W_m^M * IRC_m \right) * 100 * W_m^T$$

MCI_out^{SS} = Material Circularity Indicator regarding the circularity generated by the product for the same system

$$MCI\_out^{SS} = \sum_{m=1}^{M} \left(MC\_out_m^{SS} * W_m^M * IRC_m \right) * 100$$

MCI_out = Material Circularity Indicator regarding the circularity generated by the product

$$MCI\_out = MCI\_out^{OS} + MCI\_out^{SS}$$

If MCI $\neq$ 0 it is possible to calculate:

η_{MC} = Material Circularity Yield of the system

$$\eta_{MC} = \frac{MCI\_out}{MCI}$$

η_{RC} quantifies the circularity performance generated by other resources compared to absorbed ones. η_{RC} calculation follows the same structure presented for η_{MC} calculation.

η_{RC} = Resource Circularity Yield of the system

$$\eta_{RC} = \frac{RCI\_out}{RCI}$$

Finally, all these indexes can be used to construct a circularity function (Φ). This way, the state of the system is evaluated both considering the circularity quantity in input (CPI) and the capacity of generating circularity in output (yield vector). The yield vector length is calculated as follows:

$$\eta_C = \sqrt{\eta_{EC}^2 + \eta_{MC}^2 + \eta_{RC}^2}$$

Subsequently, the circularity function is calculated as follows:

$$\phi = \left\{ \left[\phi * CPI^2 \right] * (1 + \eta_C) \right\}$$

Considering CPI as the radius of the base circumference of a cylinder whose height is $(1 + \eta_C)$, the circularity function is equal to the volume of this cylinder. The circularity level of the system is given by CPI, but the higher the yield, the more the circularity function will grow:

- If $CPI = 0$, η_C does not exist So, $\phi = 0$
- If $CPI \neq 0$ and $\eta_C = 0$, $\phi = \left\{ \left[\pi * CPI^2 \right] \right\}$
- If $CPI \neq 0$ and $\eta_C \neq 0$, $\phi = \left\{ \left[\pi * CPI^2 \right] * (1 + \eta_C) \right\}$.

3.4 Conclusions

This chapter described a quantitative analysis model focusing on the product and calculating its circularity level. The proposed CEPA methodology is composed by three different sub-methodologies (CPA, CCA and CEA). The first one has been presented in detail, while the others have been only mentioned in a qualitative manner. Specifically, CPA allows to calculate circular shares of resource flows used in a product lifecycle, in order to obtain an exhaustive index regarding the circular percentage share of the product compared to the total resources used. Since CPA is linked with technological peculiarities and resources used for making a generic product, it is useful to compare the three different CBMs detected in Chap. 22. Outputs consist in a set of KPIs about resources circularity levels and the quantification of their economic and environmental benefits.

References

1. Brunner, P. H., & Rechberger, H. (2004). *Practical handbook of material flow analysis*. Lewis Publishers, CRC Press.
2. Hinterberger, F., Giljum, S., & Hammer, M. (2003). Material flow accounting and analysis (MFA). In *Internet Encyclopedia of ecological economics* (Issue 2).

3. Rigamonti, L., Grosso, M., & Sunseri, M. C. (2009). Influence of assumptions about selection and recycling efficiencies on the LCA of integrated waste management systems. *International Journal of Life Cycle Assessment, 14*, 411–419. https://doi.org/10.1007/s11367-009-0095-3
4. Sassanelli, C., Rosa, P., Rocca, R., & Terzi, S. (2019). Circular economy performance assessment methods: A systematic literature review. *Journal of Cleaner Production, 229*, 440–453. https://doi.org/10.1016/j.jclepro.2019.05.019
5. The Ellen MacArthur Foundation. (2015). *Towards a circular economy: Business rationale for an accelerated transition.* April 03, 2012.
6. The International Standards Organisation. (2006). *ISO 14040:2006—Environmental management, life cycle assessment, principles and framework.* https://doi.org/10.1136/bmj.332.7550.1107
7. The International Standards Organisation. (2012). *ISO/TR 14047:2012—Environmental management, life cycle assessment—Illustrative examples on how to apply ISO 14044 to impact assessment situations.*
8. Villalba, G., Segarra, M., Fernández, A. I., Chimenos, J. M., & Espiell, F. (2002). A proposal for quantifying the recyclability of materials. *Resources, Conservation and Recycling, 37*(1), 39–53. https://doi.org/10.1016/S0921-3449(02)00056-3

Open Access This chapter is licensed under the terms of the Creative Commons Attribution 4.0 International License (http://creativecommons.org/licenses/by/4.0/), which permits use, sharing, adaptation, distribution and reproduction in any medium or format, as long as you give appropriate credit to the original author(s) and the source, provide a link to the Creative Commons license and indicate if changes were made.

The images or other third party material in this chapter are included in the chapter's Creative Commons license, unless indicated otherwise in a credit line to the material. If material is not included in the chapter's Creative Commons license and your intended use is not permitted by statutory regulation or exceeds the permitted use, you will need to obtain permission directly from the copyright holder.

Chapter 4
Semi-automated PCB Disassembly Station

Simone Galparoli, Andrea Caielli, Paolo Rosa, and Sergio Terzi

Abstract The main aim of the FENIX project is the development of new business models and industrial strategies for three novel supply chains in order to enable value-added product-services. Through a set of success stories coming from the application of circular economy principles in different industrial sectors, FENIX wants to demonstrate in practice the real benefits coming from its adoption. In addition, Key Enabling Technologies (KETs) will be integrated within the selected processes to improve the efficient recovery of secondary resources. In this sense, among the available KETs, the adoption of digital and advanced automated solutions allows companies to re-thinking their business strategies, trying to cope with even more severe environmental requirements. Among these technological solutions, the paradigm of Industry 4.0 (I4.0) is the most popular. I4.0 entails the development of a new concept of economic policy based on high-tech strategies and internet-connected technologies allowing the creation of added-value for organizations and society. Unlike the activities developed in T3.1, related to the development and implementation of simulation tools and models for the smartphones' disassembly process optimization, here the attention has been spent in managing and optimizing a new semi-automated PCBs disassembly station. The disassembly of products is a key process in the treatment of Waste Electrical and Electronic Equipment. When performed efficiently, it enables the maximization of resources re-usage and a minimization of pollution. Within the I4.0 paradigm, collaborative robots (co-bots in short) can safely interact with humans and learn from them. This flexibility makes them suitable for supporting current CE practices, especially during disassembly and remanufacturing operations. D3.2 focuses on describing the semi-automated PCB disassembly process implemented at the POLIMI's Industry 4.0 Lab, aiming to demonstrate in practice the benefits of exploiting I4.0 technologies in PCB disassembly processes. Results highlight how a semi-automated cell where operators and cobots works together can allow a

S. Galparoli (✉) · A. Caielli · P. Rosa · S. Terzi
Department of Management, Economics and Industrial Engineering, Politecnico di Milano,
Piazza L. da Vinci 32, 20133 Milan, Italy
e-mail: simone.galparoli@polimi.it

© The Author(s) 2021
P. Rosa and S. Terzi (eds.), *New Business Models for the Reuse of Secondary Resources from WEEEs*, PoliMI SpringerBriefs,
https://doi.org/10.1007/978-3-030-74886-9_4

better management of both repetitive and specific activities, the safe interaction of cobots with operators and the simple management of the high variability related with different kinds of PCBs.

4.1 State of the Art on WEEE Disassembly Processes

This section presents a brief literature analysis related to relevant studies describing the relation between I4.0 and disassembly processes. The attention was placed on the role of digital technologies and techniques (specifically of autonomous robots) in supporting human operators during the disassembly process. A summary of the main contributions analysed is reported in the following sub-sections.

4.1.1 Cobots and Disassembly Processes

Since many decades, robots are exploited by manufacturers for managing complex assignments. However, common robots do not allow the direct interaction with human operators. Recently, a new generation of robots have been equipped with lots of sensors able to check the presence of humans and react accordingly during operation processes. This innovation made them more autonomous, flexible and cooperative. Under an I4.0 vision, they can interact and work safely side by side with humans and learn from them. Their flexibility enables their exploitation also for disassembly processes. Here, van den [7] describes as advances in robotics allow manufacturers to employ robots in an increasing number of applications, by increasing yield and reducing waste, as well as extending product lifecycles. Some tentative applications in industry have been described in literature. Axelsson [1] examined how to facilitate the design of a collaborative disassembly workstation in the automotive industry by means of simulation. Huang [4] evidenced as human–robot collaborative disassembly is an approach: (i) designed to mitigate the effects of uncertainties associated with the condition of end-of-life products returned for remanufacturing, (ii) able to handle unpredictability in the frequency and numbers of such returns and (iii) able to manage the variance in the remanufacturing process. Here, the authors presented a new method for disassembling press-fitted components in the automotive sector by exploiting human–robot collaboration based on the active compliance provided by a cobot. Cheng [3] focused on the assessment and optimization for manufacturing capability of human–robot collaborative disassembly to realize the aggregation and optimization of disassembly services. Here, historical data and real-time data have been fused through manifold algorithm (combining PCA, Grey correlation degree method and AHP) to get more accurate results. Liu [5] described a human–robot collaborative disassembly (HRCD) process exploiting Cyber-Physical Systems (CPSs) and Artificial Intelligence (AI), by evidencing the high feasibility and effectiveness of the cobot-based system in interacting with human operators. References [8, 9] described a cobot-assisted disassembly process dedicated to Electric Vehicle (EV) batteries. Here, humans perform complex tasks and cobots perform simple (repetitive) tasks (e.g. removing bolts and screws).

4.1.2 Cobots and WEEE Disassembly Processes

In terms of cobot-assisted WEEE disassembly processes, there are very few contributions from the literature. From one side, Bogue [2] evidenced as robotic systems exploiting artificial intelligence combined with various sensing and machine vision technologies are playing a growing role in the sorting of e-wastes, prior to recycling. To this aim, the author describes two systems for the disassembly of electronic products based on both robots and cobots, by evidencing how significant technological challenges are limiting the number of systems developed for product disassembly. From another side, Papadopoulos et al. [6] proposed a new hybrid human–robot WEEE recycling plant. Here, automatic (robotic based) systems have been exploited to categorize, disassemble, and sort products and components. What emerges from these studies is that, given the high variability of WEEEs to be disassembled, hard automated systems cannot be adopted in this context.

4.1.3 Cobots and PCB Disassembly Processes

Another important context where cobots could support human operators is in PCB disassembly and desoldering processes. However, no articles have been found in literature assessing this issue.

4.2 The Semi-automated PCB Disassembly Station at POLIMI's Industry 4.0 Lab

This section introduces the semi-automated PCB disassembly station implemented at POLIMI's Industry 4.0 Lab, a laboratory of the Manufacturing Group of POLIMI's Department of Management, Economics and Industrial Engineering completely dedicated to Industry 4.0 technologies and research. Considering existing gaps in literature, the intent of POLIMI within the FENIX project has been to exploit cobots during a PCB disassembly process in order to desolder both hazardous and valuable components from the board. This way, material recovery performances related to the subsequent hydrometallurgical recycling process could be increased, together with the overall profitability of the pilot plant. Within this section, a first overview of the developed PCB disassembly process will be presented. Then, some more details will be discussed about the operational process (and related software). Subsequently, two human-cobot collaborative interfaces developed within the FENIX project will be described. Finally, some details about the data gathering process (and the connection with Chap. 10) will be presented.

4.2.1 Structure of the PCB Disassembly Station

The semi-automated PCB disassembly station at I4.0Lab is constituted by a small amount of equipments, as: (i) a table supporting the full process, (ii) a cobot, (iii) a frame supporting PCBs during the disassembly process, (iv) a heating equipment needed to desolder components from the board, (v) a keyboard allowing the direct interaction of the human operator with the cobot, (vi) a desktop PC for the configuration of the disassembly process and (vii) an extractor hood intercepting (eventually) hazardous fumes. Figure 4.1 shows all the equipments exploited by POLIMI to implement the disassembly station within its I4.0Lab and the real configuration of the disassembly station.

4.2.2 The PCB Disassembly Process in Detail

The semi-automated PCB disassembly process starts from a clean worktop, where all the tools needed for operating have been previously positioned (see Fig. 4.2). Before proceeding with the process, the operator dresses all the personal protection devices required by safety regulations.

Once the workplace is ready to operate, the operator selects the PCB to be disassembled and the disassembly process can start.

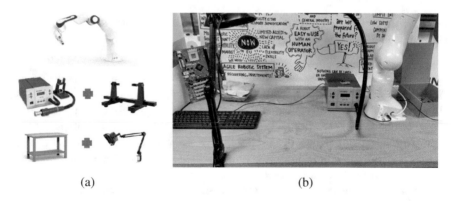

(a) (b)

Fig. 4.1 The semi-automated PCB disassembly station—**a** main equipments, **b** real configuration

Fig. 4.2 Workplace setup—**a** clean worktop, **b** tools positioning

Fig. 4.3 Front setup—**a** PCB positioning, **b** cobot arm positioning

4.2.3 Front/Back PCB Disassembly Process Setup

Subsequently, the selected PCB is mounted on the frame in order to avoid any involuntary movement during the disassembly process. The PCB is mounted with the front side directed to the cobot. Then, the cobot arm is put in the starting position in front of the PCB (see Fig. 4.3). The distance between the PCB and the cobot arm depends from: (i) the type of PCB to be disassembled, (ii) the type of component to be desoldered, (iii) the air temperature and (iv) the airflow. Finally, the operator turns on the desoldering tool and the PCB disassembly process can start.

4.2.4 Front/back PCB Desoldering Process

After the setup of the disassembly process, the operator can start working on the PCB. First, the hot airflow coming from the heating equipment is directed (through the cobot arm) to the component to be desoldered. Then, the operator (through a metal spatula) can remove all the chipsets whose pins have been already loosened by the desoldering process (see Fig. 4.4). In order to proceed with the back desoldering, the PCB is reversed on the frame and the cobot arm is reset to the starting position.

(a)   (b)

(c)   (d)

Fig. 4.4 Front desoldering—**a** initialization, **b** desoldering 1, **c** desoldering 2, **d** finalization

At the end of the PCB disassembly process (see Fig. 4.5), all components are desoldered from the board and both components and board can be sent to UNIVAQ's hydrometallurgical recycling process for material recovery.

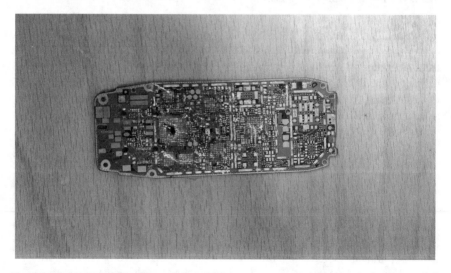

Fig. 4.5 Final PCB desoldering process result

4.3   ROS-Based Control Architecture Setup

The robot controller can be divided into two levels. The low-level control and the high-level control. The whole control application is built using ROS (Robot Operating System) to grant a standardized open architecture with the possibility of adding new component interacting with the existing ones in an easy and open way.

The setup is composed of several modules. Each of them takes care of a specific functionality using the ROS topic-based communication architecture to share services and information with other modules. Together with ROS_Kinetic as core element, other third-party essential modules and libraries are included, like:

- LibFranka Is a C++ library developed by Franka® Emika, the cobot producers
- FrankaRos Is a collection of ROS packages developed from Franka® Emika
- Moveit is a ROS motion planning framework.

The main nodes developed for the control architecture are:

- Fenix main: Manages all the Moveit entities and provides services for moving the cobot
- Fenix manager: Manages the logic states of the cobot, tracks the actual state and defines the target pose
- Keyboard interface: Provides a simple interface to command the cobot from keyboard keys.

4.3.1   Low-Level Real Time Controller

The low-level real time controller is based mainly on three parts.

The first one is the ROS controller that takes as input a trajectory, compute, and communicate the required joint speed to the Franka® Control Interface (FCI). The communication takes place over ethernet and ensures 1 kHz control loop between the control software and the cobot Hardware.

The Franka® Control Interface inside the cobot is responsible of the actuation of the command received from the control unit and of all the safety check described below.

The Loop is closed from the FCI with a constant flow of information on the kinematical and dynamical parameters of the cobot itself forwarded to the ROS controlled and available inside the ROS High-level architecture.

The low-level safety measures are based on torque sensors present on each joint of the cobot. The sensors allow the FCI to monitor the force that each link is applying at every time instant. To operate the cobot needs as information the weight of the loads attached to the end effector. With this parameter it can calculate the theoretical torque each joint need to produce in order to grant the required kinematic and dynamic state. If the measured values differ from the estimated ones more than the specified tolerance the cobot has detected an undesired contact and, in the actual configuration, interrupt its operation instantaneously waiting for a feedback from the operator in order to resume the work.

Other strategies that can be implemented are based on a dynamic response to unwanted interaction such as elastic behaviour or dynamic trajectory re-planning.

4.3.2   High-Level Task and Safety Management Controller

The high-level control is in charge to define, based on a set of input, the desired position for the cobot end effector, and the desoldering tool. In the actual configuration the input is keyboard based and is represented by a couple of direction and distance representing a vector in 3D space but aligned with one of the three axes. The Control unit sums this vector with the actual position and determines the next desired position. The orientation for the desired pose is considered fixed and with the desoldering tool pointing directly to the PCB. This is the most efficient working position and, has been tested, the easiest way for an operator to clearly visualize and instruct the robot on where to place itself.

From the safety aspect together with the automatic collision detection system discussed above some additional safety measures are included.

From a physical point of view, the operator can use three different emergencies stop procedures.

The first one is associated to a key on the keyboard that tells the robot management system to send the command to stop the cobot. This procedure allows the cobot to immediately restart working by pressing another key and is the lightest emergency stop mode.

The second is a physical separated button that tells directly to the FCI to stop the cobot in case of high-level control failure. After this procedure, the cobot should be re-initialized to start operating again.

The third is a power cut-off button that physically interrupt the power supply to all the motor and stops them. With the power cut off the cobot will collapse since is maintained in it is position by the motors. For this reason, each joint possesses a pneumatic emergency stopper. During the normal operation, the pneumatic systems keeps the stoppers open and the cobot can move but when the power is off the stoppers closed and the arm is fixed in position preventing further movement, operator injures and instrumentation damages.

The last safety measure is the decision to keep the maximal movement speed for the end effector very low (below 10 cm/s) in order to test the interaction between Human operator and cobot in the safest condition as possible, even if, it's clear that this is a limit in the actual performances.

4.3.3   Cobot State and Trajectory Planning Real Time Visualization Tool

The cobot calculate the desired pose from a direct operator input but the way to reach the desired pose perform some more sophisticated calculation. The first aspect to underline is that the cobot possess a knowledge of the 3D spaces around itself, given on startup and therefore its operative space is restricted to avoid collision and injures. The trajectory is calculated in order online for each movement with the OMPL planner using RRT* algorithm.

The trajectory planning wants to minimize:

- The deviation of the tool from the predetermined work plane to increase the performance of the operation.
- The deviation from the operator's expected trajectory to help the human operator understanding the cobot behaviour.

Together with the trajectory plan the operational space and the planned trajectory are visualized in RViz (Fig. 4.6) in order to allows the operators an even better understanding of the cobot actual state and give a clear feedback.

The 3D real time robot visualizations allow also to disconnect the robot controller and connect the trajectory planner only to the RViz simulation. This is useful in case

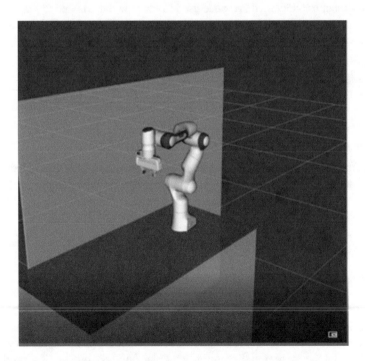

Fig. 4.6   The RViz real time cobot visualization

of test or operators training, since the input are executed in the virtual operating space from the virtual cobot based on the same logic that in a real situation but in a safer environment, providing a clear real time feedback.

4.3.4 Operator-Oriented Manual Control Interface

The manual control interface has been developed to give the operator a way to operate the cobot that is easy and intuitive to allows him to focus on the desoldering operation. Since the orientation of the end effector is decided from the cobot autonomously (and in the first setup is fixed) the operator needs only to input the desired movement of the end effector within respect with the current position.

This can be easily done by an arrow key-based interface. The operator can choose with two keys (" + " and "–") the amount of cm (in step of 5 mm) he wants the cobot to move and with 6 keys (2 for each axes) the direction of the movement.

4.3.5 Real Time Process Data Gathering Tool

The last aspect implemented is a node used to save the data related to the cobot and desolder tool's state together with the current state of the desoldering operation.

Each second the following data are saved:

- Time
- Sample ID
- Operational state
- Energy Consumption
- Cobot Joint position
- Cobot Joint Acceleration.

The data are used to perform an energetic analysis of the operation together with a performance analysis to guide the future improvement of the Setup.

4.3.6 Desk Web Interface

Panda Desk is a web interface published by the cobot's producer.

Panda offers the easiest and fastest workflow-based user experience. Using Desk—Franka® Emika's browser-based interface—apps can be arranged to create entire tasks in no time. These tasks can quickly be adapted, reused, or deployed on multiple robots to remarkably reduce setup costs. Robot apps incorporate the entire complexity of the system and represent modular building blocks of a production process such as grasping, plugging, insertion and screwing. Individual apps can be parameterized by means of showing Panda poses by demonstration, or adding context relevant parameters such as speed, duration, forces, and triggering actions.

4.3.7   FRANKA® Control Interface

The Franka® Control Interface (FCI) allows a fast and direct low-level bidirectional connection to the Arm and Hand. By using libfranka, an open source C++ interface, you can send real-time control values at 1 kHz. In addition, rankaROS connects Franka® Emika research robots with the entire ROS ecosystem. It integrates libfranka into ROS Control. Additionally, it includes URDF models and detailed 3D meshes of the robots and end effectors, which allows visualization (e.g. RViz) and kinematic simulations. MoveIt! integration allows pose and trajectory plan and execution.

4.4   Application and Results

4.4.1   Manual Desoldering Tests

The first tests conducted without the usage of the cobot aims to confront the results of the Literature research related to the desoldering operations with a real operation in a real environment and with a real equipment.

The tests conducted have allowed us to define operation parameter for the cobot-assisted procedure. The main involved parameters are 4:

- Temperature of hot air
- Flow power of hot air
- Time to desoldering different components
- Distance of the tools from the components.

Together with the parameter the tests aim to test the safety equipment provided to the operator to grant his safety.

4.4.2   Cobot-Assisted Desoldering Tests

The tests have been conducted on two sets of smartphone's PCBs with similar characteristics. The differences present in the test set represent the real industrial condition since smartphones PCB are an extremely variable product.

The first set of PCBs has been used to perform some qualitative tests to detect problem in the setup and gather feedback from the operators. For this reason, half of the PCBs have been processed by one of the developers of the application and the other half by a new operator trained during the process from the expert one.

The feedback gathered helped in the development of the keyboard interface, of the data acquisition module and in the definition of the rues for the trajectory planning in order to give the correct feedback to the operator in terms of response time and expected behaviour.

The second set of 50 PCBs has been disassembled to test the performance of the setup following the steps here described:

- The PCB is weighted
- The PCB is disassembled, and the process data are gathered automatically
- The PCB is weighted again to detect the removed components' weight
- The operator reports any problem during the operation.

The data obtained from this set of tests are the ones used to produce the reports of the experiments.

4.4.3   Data Gathering from Cobot-Assisted Desoldering Tests

The data gathering infrastructure described above presents one main critical aspect. The electrical consumptions for the cobot and for the desoldering tool are estimated based on the state of the equipment and on the nominal consumption data provided from the producer. This limits the real analysis capability and is, therefore, one of the aspects we aim to improve in the next activities with the cobot, even if not as part of the FENIX project.

The solution for this problem is clearly the installation of a smart power measurement system capable of automatically deliver the measured data to ROS in order to be integrated in the data generated ad described in Chap. 10.

The sensors will directly monitor the power consumption from the electrical network to generate data with a frequency of 2 Hz. Each data is an average of the consumption in the past 0.5 s.

The data are published through MQTT or can be obtained by issuing a rest call to the device itself.

The Presence of two sensors one for each of the two power-consuming assets will give us a clear understanding of where the most critical power consumption is and clues on how to reduce the energetic impact of the operation.

4.5 Conclusions

This chapter described the implementation a semi-automated PCB disassembly process carried out at the POLIMI's Industry 4.0 Lab. Considering the results, the cobot-assisted desoldering process allows to completely and easily desolder components from PCBs in about 9 min/unit. However, after some preliminary estimates, the full operation time could be furtherly reduced in about 4 min/unit. As already discussed in other sections, this experiment allowed to demonstrate in practice what should be the benefits and performances reachable through the introduction of I4.0 technologies in current WEEE management processes. Here, the role of a human operator is essential for two reasons: (i) the fast identification of valuable components to be removed and (ii) the fast decision-making and problem solving during too complex activities for the cobot.

References

1. Axelsson, J. (2019). *Using IPS software for decision making when developing a collaborative work station—A simulation-based case study in the remanufacturing industry.* Malardalen University, Sweden.
2. Bogue, R. (2019). Robots in recycling and disassembly. *Industrial Robot, 46*(4), 461–466. https://doi.org/10.1108/IR-03-2019-0053
3. Cheng, H., Xu, W., Ai, Q., Liu, Q., Zhou, Z., & Pham, D. T. (2017). Manufacturing capability assessment for human-robot collaborative disassembly based on multi-data fusion. *Procedia Manufacturing, 10*, 26–36. https://doi.org/10.1016/j.promfg.2017.07.008
4. Huang, J., Pham, D. T., Wang, Y., Qu, M., Ji, C., Su, S., Xu, W., Liu, Q., & Zhou, Z. (2019). A case study in human–robot collaboration in the disassembly of press-fitted components. *Proceedings of the Institution of Mechanical Engineers, Part B: Journal of Engineering Manufacture*, 1–11. https://doi.org/10.1177/0954405419883060
5. Liu, Q., Liu, Z., Xu, W., Tang, Q., Zhou, Z., & Pham, D. T. (2019). Human-robot collaboration in disassembly for sustainable manufacturing. *International Journal of Production Research, 57*(12), 4027–4044. https://doi.org/10.1080/00207543.2019.1578906
6. Papadopoulos, G. T., Axenopoulos, A., Giakoumis, D., Kostavelis, I., Papadimitriou, A., Wollherr, D., Sillaurren, S., Bastida, L., Oguz, O. S., Wollherr, D., Garnica, E., Vouloutsi, V., Verschure, P. F. M. J., Tzovaras, D., & Daras, P. (2019). A hybrid human-robot collaborative environment for recycling electrical and electronic equipment. *SmartWorld 2019-ubiquitous intelligence & computing, advanced & trusted computing, scalable computing & communications, cloud & big data computing, internet of people and smart city innovation (SmartWorld/SCALCOM/UIC/ATC/CBDCom/IOP/SCI)*, pp. 1754–1759. https://doi.org/10.1109/SmartWorld-UIC-ATC-SCALCOM-IOP-SCI.2019.00312
7. van den Beukel, J.-W. (2017). *Industry 4.0 as an enabler of the circular economy: Preventing the waste of value and permitting the recovery of value from waste.* PWC Sustainability and Climate Change Blog. https://pwc.blogs.com/sustainability/2017/06/industry-40-as-an-enabler-of-the-circular-economy.html
8. Wegener, K., Andrew, S., Raatz, A., Dröder, K., & Herrmann, C. (2014). Disassembly of electric vehicle batteries using the example of the Audi Q5 hybrid system. *Procedia CIRP, 23*, 155–160. https://doi.org/10.1016/j.procir.2014.10.098

9. Wegener, K., Chen, W. H., Dietrich, F., Dröder, K., & Kara, S. (2015). Robot assisted disassembly
    for the recycling of electric vehicle batteries. *Procedia CIRP, 29*, 716–721. https://doi.org/10.
    1016/j.procir.2015.02.051

Open Access This chapter is licensed under the terms of the Creative Commons Attribution 4.0
International License (http://creativecommons.org/licenses/by/4.0/), which permits use, sharing,
adaptation, distribution and reproduction in any medium or format, as long as you give appropriate
credit to the original author(s) and the source, provide a link to the Creative Commons license and
indicate if changes were made.

The images or other third party material in this chapter are included in the chapter's Creative
Commons license, unless indicated otherwise in a credit line to the material. If material is not
included in the chapter's Creative Commons license and your intended use is not permitted by
statutory regulation or exceeds the permitted use, you will need to obtain permission directly from
the copyright holder.

Chapter 5
A Mobile Pilot Plant for the Recovery of Precious and Critical Raw Materials

Ionela Birloaga, Nicolo Maria Ippolito, and Francesco Vegliò

Abstract In order to furtherly proceed with the recycling of raw materials from e-wastes, PCBs must be treated in a hydrometallurgical process able to extract useful materials from them. This chapter presents some details of the hydrometallurgical pilot plant developed in FENIX.

5.1 Introduction

Precious metals (PMs) are crucial in the global economy as they are key constituents of a vast number of industrial products and processes. Large amounts of wastes with various contents of precious metals are generated every year. The wastes of electrical and electronic equipment know the worldwide largest increment. With the current growing trends there is estimated that this amount will arrive to 120 metric tons/year by 2050 and the consumption of the raw materials with be two-fold [1]. The waste printed circuit boards represent an important secondary resource of precious metals (Au, Ag, Pd) but also of base metals (Cu, Zn, Ni, Sn, Pb, Fe, Al). As was expressed in the paper of Wand and Gaustad [2] the main economic drivers in the recycling of such waste, considering their concentrations and market price, are in the following order: Au, Pd, Cu, Ag, Pt, Sn and Ni. According to the study of Golev et al. [3], the waste printed circuit boards represent about 40% of the metal recovery value from the entirely equipment.

The application of hydrometallurgical methods for recycling is preferred to pyrometallurgical methods, as the latter usually require high temperatures, produce harmful gases (such as SO_2) and dust, and require high capital costs.

The first step of each hydrometallurgical recycling technique is represented by leaching. In order to achieve a high leaching efficiency of precious metals (PMs), the aqua regia, cyanide, thiol groups (thiourea, thiosulfate and thiocyanate), halides

I. Birloaga (✉) · N. M. Ippolito · F. Vegliò
Department of Industrial Engineering, Information and Economy, University of L'Aquila,
Monteluco di Roio, 67100 L'Aquila, Italy
e-mail: ionelapoenita.birloaga@univaq.it

© The Author(s) 2021
P. Rosa and S. Terzi (eds.), *New Business Models for the Reuse of Secondary Resources from WEEEs*, PoliMI SpringerBriefs,
https://doi.org/10.1007/978-3-030-74886-9_5

(chloride, iodide and bromide) with the presence of different oxidants (oxygen, ferric complexes, hydrogen peroxide, chlorine, bromide, iodine) are employed. The generations of highly polluted NOx and HCN gases, as well as various harmful elements in wastewater, have restricted the use of aqua regia and cyanide leaching systems. Table 5.1 presents a patent list of the hydrometallurgical processes with their brief overview that are used for base and precious metals recovery from WPCBs.

The hydrometallurgical processes have gained the largest interest of application for waste printed circuit boards treatment. In addition the circular economy principle stared to be of high interest for researchers that activate in the field of WPCBs treatment. However, till present, for recovery of all elements that are present within the structure of WPCBs was not possible to be obtained using hydrometallurgical processes.

For FENIX Project, the authors of this chapter have developed and tested the efficiency of two hydrometallurgical technologies for e-waste recycling at both laboratory and pilot levels. For both processes, commercially named GOLD REC 1 [9] and GOLD REC 2 [10], patent applications have been deposited at both Italian and European levels. The main core was to recover both precious and base metals content from electronic waste and to use them as material for the production of metallic powders (USE CASE 1), 3D printing filaments (USE CASE 3) and jewelries (USE CASE 2). Within this chapter the processes description and a summary of the activities undertaken for this core achievement are presented.

5.2 Pilot Plant Design and Description by Process Performing

The pilot plant was designed considering its installation in a real industrial environment logistically useful to carry out experimental tests for the researchers involved in this project. The advantages to operate in this industrial site are:

- Working in a real environmental context;
- Availability of the e-waste necessary for the pilot plant;
- Availability of several services (compressed air, grinding section, electricity, working men for dismantling, grinding and other technical operations)

In this way the following action were carried out:

a. Design and construction of the pilot plant;
b. Testing activities (comparing in parallel the same results with pilot lab-scale tests)
c. Production of some suitable amount of materials for their characterization and to be used for the other partners of the FENIX project.

The pilot plant was constructed within a container that was devised within three sections, namely: one section for operator and control panel, a second section that

Table 5.1 Patents on waste printed circuit board treatment for recovery of both base and precious metals by hydrometallurgical procedures

References	Description	Remarks
[4]	This patent presents the following steps for recovery of Cu, Au, Ag, Pd, Pt, P and Sn from waste printed circuit boards: – Crushing and gravity separation to achieve a Cu concentrate and removal of non-metallic part – Smelting of the Cu concentrate and its electro refining in a solution of $CuSO_4$ and H_2SO_4 (a 4 N copper ingot was achieved) – Recovery of anode slime and treatment with NaCl, H_2SO_4 and $NaClO_3$ to selectively recover gold, palladium and platinum. For the gold reduction from solution, Na_2SO_3 was used. Then, the solution was neutralized to pH 2 and zinc and iron powders were used for Pd and Pt displacement – The resulted sludge after Au, Pd and Pt leaching, was further leached with Na_2SO_3 to recover silver. Afterwards, this was reduced with oxalic acid or formaldehyde – The leaching solid residue was then heated in the mixed solution of HCl, NaCl and $CaCl_2$ to recover lead. After residual solid separation by filtration, the solution was cooled favouring within this way Pb precipitation. Thereafter, the solution regeneration was performed with calcium chloride and then reutilized within Pb leaching process – The achieved solid residue was then roasted in the presence of sodium hydroxide and sand to recover Sn. Then, the roasting product was mixed with water and the solid residue was filtered out. To achieve the Sn recovery as Na_2SnO_3, the solution of was evaporated and crystallized	The authors have developed a pyro-hydrometallurgical process that was applied on waste printed circuits boards after a physical-mechanical pre-treatment. This procedure consists in crushing of the boards and then non-metallic parts separation by the metallic ones by air classification. However no data on the efficiency of this technology have been provided within this invention. Generally this process do not allows the complete separation of nonmetals and in addition, by performing the smelting process, in case of nonmetals presence, toxic gases are produced. The authors sustain that all the waste solutions have been recirculated within the process

(continued)

Table 5.1 (continued)

References	Description	Remarks
[5]	The high grade printed circuit boards have been treated in sulfuric acid and hydrogen peroxide solution and then the detached gold fingers have been treated by two ways: (i) leaching with diluted nitric acid for Cu removal and then the purified gold is smelted and (ii) dissolution of gold fingers within aqua regia at a certain temperature, cooling of solution, filtration, neutralization to pH 1, addition of zinc particles for Au cementation and then, to remove the excess of zinc, the treatment of the precipitate with diluted hydrochloric acid at certain temperature is performed. The resulted solution after copper leaching with sulfuric acid and hydrogen peroxide is concentrated by evaporation and $CuSO_4$ crystals are achieved	The current invention presents different systems of reagents for Cu and Au recovery. No discussion regarding the waste solutions treatment or other elements recovery is shown
[6]	Within the current invention the researchers have firstly performed the roasting of the boards, then the leaching with sulfuric acid to extract the base metals and then the solid residue leaching with aqua regia at a certain range of temperature. Thereafter, the nitric acid was completely removed by maintaining a certain level of temperature. The achieved solution was then used for selective precipitation of Ag, Pd and Au	Within this invention, the recovery of precious metals after a thermal treatment and base metals removal by acid leaching was performed. Aqua regia was also used within this patent for Au, Ag and Pd dissolution and then, after selective precipitation of all three precious metals, the recirculation of solution was achieved
[7]	The invention has as core to recover Au from minerals by leaching using an iodine salt in presence of a carboxylic acid (citric and acetic acid), boric acid and optionally chlorine or bromine. The leaching is performed on fine materials in an electrochemical cell which allows the increase of the ORP to oxidize the iodide to iodine. Moreover, the inventors have present as suitable alternatives the addition of other oxidants (H_2O_2, NaOCl, O_3, etc.) that are capable to perform the oxidation	The advantage of this process is the fact that the solution can be regenerated and therefore used many times. However, the authors of this patent have not presented any data on the precious metals recovery from solution. In addition, there are also other elements that can be leached within this media. This fact is not pointed out within this patent

(continued)

Table 5.1   (continued)

References	Description	Remarks
[8]	This patent presents the recovery of precious and base metals from waste printed circuit boards using the following operations: – Stripping process with concentrated nitric acid, ferric chloride and ferric nitrate to recover Sn and Pb and the electronic components – Shredding of the boards, eddy current separation to achieve a high concentrate of metals, treatment of the concentrate with sulfuric acid and an oxidant to recover Cu and Ni and then electrolysis to achieve Cu and Ni metals – Dissolution of silver and palladium from the solid residue of the previous leaching process using nitric acid. Then, by neutralization with NaOH till pH 3–5, the palladium recovery from solution was achieved. Silver was recovered by addition of a chloride salt – The solid residue of previous step was then leached with aqua regia for Au leaching. The resulted solution was further subjected to electrowinnig process to recover metallic gold	The current invention allows the recovery of both precious and base metals from WPCBs using various reagents. However, due to the fact that the patent is in Chinese language, was difficult to understand well the entirely process

has a chemical reactor (R101) with an useful volume of 200 L and a working temperature of 70 °C; this reactor was and it is used for leaching, precipitation and cementation; a filter press (FP 101) for filtration of the solutions that have more than 1 g/L of solid content (e.g. solid residue of leaching or base metals precipitates) and a candle filter for solutions with a solid content lower than 1 g/L; two electrochemical cells (EC1—for base metals and EC2—for precious metals); one scrubber; the third section with 12 storing tanks (TK 101–106 for reagents and TK 107–112 for solutions and wastewater) (See Figs. 5.1, 5.2 and 5.3).

5.2.1   GOLD REC 1 Process Description

The hydrometallurgical process has started with the HydroWEEE EU Project and fully developed within the HydroWEEE DEMO EU Project. The current hydrometallurgical procedure, as is shown in Fig. 5.3, consists in the following operations:

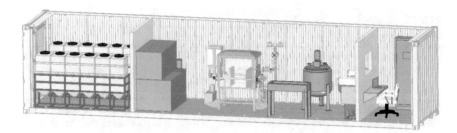

Fig. 5.1 FENIX hydrometallurgical plant—3D view

- The waste printed circuit boards are firstly subjected to a physical mechanical procedure where the Al- and Fe-based components are removed from PCBs surface. Then, the depopulated PCBs are shredded and milled to suitable particles sizes;
- The milled PCBs are then leached with water, sulfuric acid and hydrogen peroxide for the extraction of base metals by the precious metals (Eqs. 5.1 and 5.2).

$$Cu + H_2SO_4 + H_2O_2 = CuSO_4 + 2H_2O \tag{5.1}$$

$$Sn + H_2SO_4 + H_2O_2 = SnSO_4 + 2H_2O \tag{5.2}$$

- The solid separation by the leach liquor is carried out by filtration process followed by washing with water. The resulted solution is subjected to a precipitation process for Sn precipitation. Then, also this solid precipitate is separated from solution by filtration and further washed with water. The solution achieved after Sn recovery is sent to an electrowinning cell for Cu recovery (Eq. 5.3).

$$CuSO_4 + H_2O + 2e^- = Cu + H_2SO_4 \tag{5.3}$$

- Then, the resulted solution is recycled in the first leaching process for the leaching of another PCB material.
- The solid residue of base metals leaching process is involved into another leaching process with thiourea as reagent, ferric sulfate as oxidant in diluted sulfuric acid for Au and Ag dissolution (Reactions 5.4 and 5.5).

$$Au + 2CS(NH_2)_2 + Fe^{3+} \rightarrow Au[CS(NH_2)_2]_2 + + Fe^{2+} \tag{5.4}$$

$$Ag + 3CS(NH_2)_2 + Fe^{3+} \rightarrow Au[CS(NH_2)_2]_3^+ + Fe^{2+} \tag{5.5}$$

- Then, after removal of solid suspension form solution by filtration, the electrowinnig process is also applied on this solution for Au and Ag recovery (Eqs. 5.6 and 5.7). Once the process is finished, the solution is also recycled for leaching

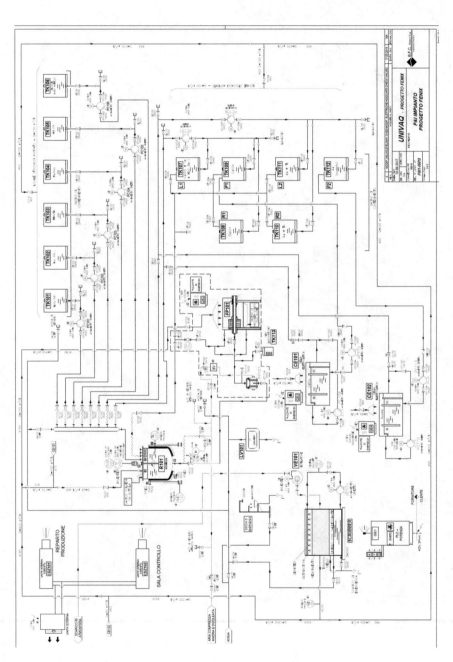

Fig. 5.2 P&I of the FENIX's hydrometallurgical plant

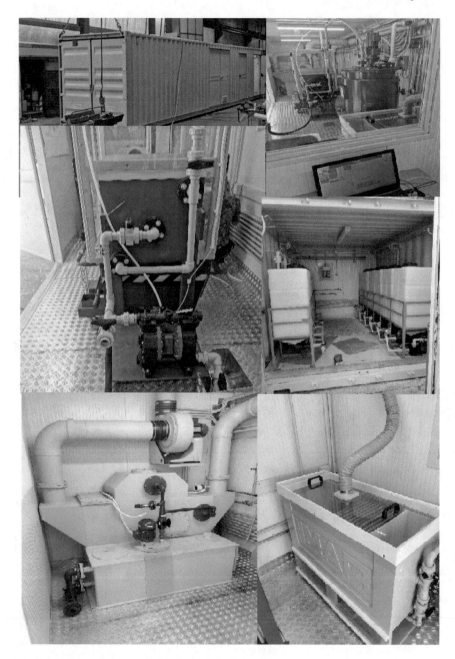

Fig. 5.3 FENIX's hydrometallurgical plant—real view

of precious metals from other solid residue of base metals leaching process.

$$Au(CSN_2H_4)_2^+ + e^- = Au^0 + 2CSN_2 \qquad (5.6)$$

$$Ag(CSN_2H_4)_3^+ + e^- = Ag^0 + 3CSN_2H_4 \qquad (5.7)$$

- It is important to specify that solutions recycling after electrowinnig process will not be total. A part of these solutions are treated by proper technologies of waste water treatment. The treatment of the wastewater coming from the base metals recovery step consist in precipitation of the impurities with calcium hydroxide. The residual solution of precious metals recovery step is treated firstly with hydrogen peroxide and ferrous sulfate for degradation of organic complexes and then with calcium hydroxide for impurities precipitation. At the end of wastewater treatment process, the filtration is performed for solid removal from the treated water (Fig. 5.4).

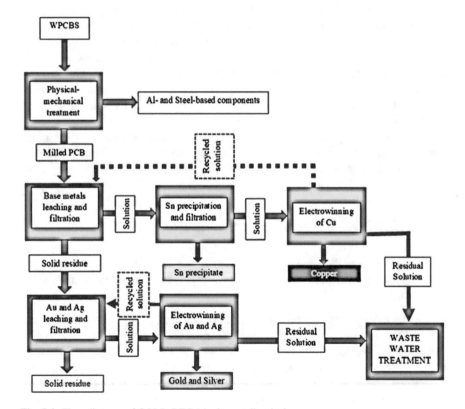

Fig. 5.4 Flow diagram of GOLD-REC 1 hydrometallurgical process

Fig. 5.5 SEM image and photography of copper deposit

Various tests of metals recovery with this hydrometallurgical technology have been applied on a milled sample of WPCBs of personal computers. Under the optimal conditions of the leaching process (solid concentration of 15%, under continuous agitation for 2 h for each step and a reagents concentration of 1.8 M of sulfuric acid and 20% vol/vol of hydrogen peroxide), which take place using the two-step counter current method, over 95% of Cu recovery and 60% for Sn have been achieved. The resulting solution was subjected to a coagulation process with polyamine solution in a concentration of 10% wt./vol. and 90% of tin content from solution was recovered. The obtained product had a tin concentration of about 50%. Furthermore, the electrolysis process was performed using graphite as cathode and zirconium-titanium electrode as anode.

At the end of the process, the purity of Cu product was 89% and the determined power consumption was 2.39 kWh/kg of Cu. The obtained copper deposit (Fig. 5.5) has been used for the additive manufacturing process (USE CASE 1).

5.2.2 GOLD REC 2 Process Description

The original process patented is presented in Fig. 5.6.

This hydrometallurgical process could be synthetically described as indicated in the follow:

- The chemical process can be applied on the e-waste without grinding (with whole WPCB as an example) avoiding important loss of precious metals also described in the literature;
- The process uses san unique step of metals dissolution with a chemical leaching using HCl, H_2O_2, acetic acid in water solution at room temperature (21 °C ± 3 °C) with a solid/liquid ratio of 10–20% (Eq. 5.8–5.15). The chloroacetic acid is produced by in situ chemical process within two steps: firstly, hydrochloric acid reacts with hydrogen peroxide and acetic acid to produce peracetic acid, water and chlorine (Eq. 5.8); in the second step chloroacetic acid and hydrochloric acid

E-WASTE

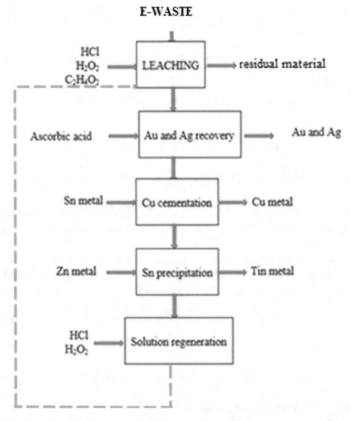

Fig. 5.6 Flow diagram of GOLD REC 2 hydrometallurgical process

are produced by the chlorination of the unreacted acetic acid (Eq. 5.9). The global reaction of this process is represented by Eq. 5.10.

$$2\,HCl + 2\,H_2O_2 + C_2H_4O_2 = C_2H_4O_3 + 3\,H_2O + Cl_2 \qquad (5.8)$$

$$C_2H_4O_4 + Cl_2 = C_2H_3ClO_2 + HCl \qquad (5.9)$$

$$HCl + H_2O_2 + C_2H_4O_2 = C_2H_3ClO_2 + 2\,H_2O \qquad (5.10)$$

$$1.5\,C_2H_3ClO_2 + 1.5\,HCl + Au = AuCl_3 + 1.5\,C_2H_4O_2 \qquad (5.11)$$

$$C_2H_3ClO_2 + HCl + 2\,Ag = 2\,AgCl + C_2H_4O_2 \qquad (5.12)$$

$$C_2H_3ClO_2 + HCl + Cu = CuCl_2 + C_2H_4O_2 \tag{5.13}$$

$$C_2H_3ClO_2 + HCl + Sn = SnCl_2 + C_2H_4O_2 \tag{5.14}$$

$$C_2H_3ClO_2 + HCl + Ni = NiCl_2 + C_2H_4O_2 \tag{5.15}$$

$$C_2H_3ClO_2 + HCl + Pb = PbCl_2 + C_2H_4O_2 \tag{5.16}$$

$$C_2H_3ClO_2 + HCl + Zn = ZnCl_2 + C_2H_4O_2 \tag{5.17}$$

- Precious (Au and Ag) and base metals (Cu, Sn, Zn, Ni, Pb) are dissolved leaving the WPCB with mainly epoxy resins and fiberglass structure intact (with some residues of metals);
- The liquid solution is easily separated from the S/L system and selective reduction-precipitations steps are considered in the process to recover the dissolved metals. These steps are synthetically described in the follow:

a. Reduction and precipitation of Au chloride to its metallic form by ascorbic acid;

$$AuCl_3 + 1.5\ C_6H_8O_6 = Au + 3HCl + 1.5C_6H_6O_6 \tag{5.18}$$

b. Cooling the solution to less than 15 °C for precipitation of AgCl;
c. Selective reduction and precipitation of Cu by metallic Sn or co-reduction of both copper and tin ions with iron metal;

$$CuCl_2 + Sn = SnCl_2 + Cu \tag{5.19}$$

$$CuCl_2 + Fe = FeCl_2 + Cu \tag{5.20}$$

$$SnCl_2 + Fe = FeCl_2 + Sn \tag{5.21}$$

d. Reduction and precipitation of SnCl$_2$ by metallic Zn;

$$SnCl_2 + Zn = ZnCl_2 + Sn \tag{5.22}$$

e. Exploitation of the residual solution for its recycling within the process or by adding iron in order to produce a $FeCl_2$-$FeCl_3$ solution useful for coagulation processes in the treatment of wastewaters;

2. The main products are: Au (after melting process in an inductive electrical oven adding some slug compound), AgCl, Cu and Sn in powder forms (mainly in the range of 10–90 μm) and a residual chloride solution that can be regenerated by make-up with proper reagents concentration or treated with iron metal to achieve a high concentrated iron solution (extensively and usually utilized in the coagulation processes in wastewater treatments);

Various tests have been conducted at both laboratory and pilot levels. These were carried out on various streams, namely: RAM modules, PCBs of mobile phones and CPU. The runs were performed using the following conditions: 3.5 M of HCl, 10% wt./vol. of $C_2H_4O_2$, 5% wt./vol. of H_2O_2, 15–20% of solid concentration, room temperature, 3 h. Under these conditions, recovery yields between 60 and 95% were achieved for Au, Ag, Cu and Sn content of the three kind of waste. Not complete dissolution is achieved since the process is performed in whole material. The waste materials (RAM modules and PCB of mobile phone) have entrapped within their layers and components these elements. Therefore, not complete exposition of elements to the leaching media is realized. The results carried out within the hydrometallurgical plant revealed a recovery of about 50–75% for these four elements. The process of reduction with ascorbic acid had an efficiency of over 95% for Au recovery at both pilot and laboratory levels. Figure 5.7 presents the photographic aspects of one of

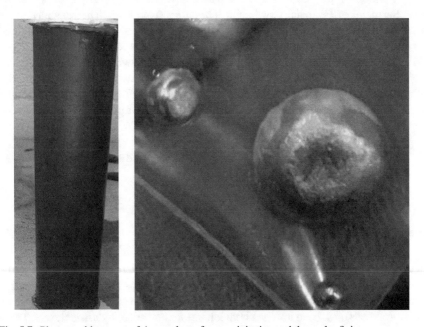

Fig. 5.7 Photographic aspect of Au product after precipitation and thermal refining

Fig. 5.8 Copper metal powder composition

the Au precipitates recovered at pilot level and final achieved product after thermal refining.

The further step of AgCl precipitation from solution by cooling revealed over 75% of recovery at laboratory scale after 3 h of reaction. At pilot level, Ag was coprecipitated during the copper cementation process. This was mainly since the plant does not have a cooling system. The copper recovery was performed either with Sn metal or Fe metal powders at laboratory scale level. The runs have been carried out at different stoichiometric amounts of both base metals and the optimal results in terms of recovery from solution and purity of products was achieved with tin metal at a stoichiometric amount of 0.8 (82% of recovery and 97% of purity). This is mainly since tin metal has close value to copper within the reactivity series of metals. The copper recovery with Fe revealed over 99% of recovery of copper at a stoichimetric excess of 45% and a purity of 84%. At pilot level, the produced copper powder (Fig. 5.8) had a copper content of about 50% with Fe and Sn as main impurities,

The achieved gold and copper products have been used for jewelry (USE CASE 2) and filaments (USE CASE 3) production.

5.3 Conclusion

In order to achieve a circular economy for metals, FENIX Project, has as one of the main cores to perform the recovery of base and precious metals from e-wastes and to reuses them for manufacturing of new products. For this reason, two hydrometallurgical processes have been tested at both laboratory and small industrial levels. There have been performed various experiments and according to the results achieved at pilot level, both technologies must be further improved to achieve better recovery degrees and properties of final products.

References

1. The World Economic Forum. (2019). *A New Circular Vision for Electronics, Time for a Global Reboot- Report* (p. 24).
2. Wang, X., & Gaustad, G. (2012). Prioritizing material recovery for end-of-life printed circuit boards. *Waste Management, 32*(10), 1903–1913.
3. Golev, A., Schmeda-Lopez, D. R., Smart, S. K., Corder, G. D., & McFarland, E. W. (2016). Where next on e-waste in Australia? *Waste Management, 58,* 348–358.
4. Zhang, S., Li, B., Pan, D., Tian, J., & Liu, B. (2012). *Complete non-cyanogens wet process for green recycling of waste printed circuit board*. US20120318681A1.
5. Zha, H., He, L., Wen, W., & Liu, C. (2014). *Process for recycling gold from waste circuit boards*. CN104152696 (A).
6. Brunori, C., Fontana, D., Carolis, R. D., Pietrantonio, M., Pucciarmati, S., Guzzinati, R., & Torelli, G. N. (2015). *Hydrometallurgy process for the recovery of materials from electronic boards*. WO2015052658A1.
7. Nelson, D., Scott, S., Doostmohammadi, M., & Jafari, H. (2017). *Methods, materials and techniques for precious metal recovery*. US20170369967A1.
8. Kaihua, X., & Li, Y. (2011). *Method for separating and recycling rare noble metals and waste plastics in waste circuit board*. CN102240663 (A).
9. WO2018215967A1 Process for the hydrometallurgical treatment of electronic boards, Inventors: Birloaga, I., Vegliò, F., De Michelis, I., & Ferella, F. (2018). Priority number IT201700057739 A·2017-05-26 (Gold-REC1)
10. WO2019229632A1, Hydrometallurgical method for the recovery of base metals and precious metals from a waste material, Inventors Birloaga, I., & Vegliò, F. (2019). Priority number - IT201800005826A·2018-05-29 (Gold-REC2)

Open Access This chapter is licensed under the terms of the Creative Commons Attribution 4.0 International License (http://creativecommons.org/licenses/by/4.0/), which permits use, sharing, adaptation, distribution and reproduction in any medium or format, as long as you give appropriate credit to the original author(s) and the source, provide a link to the Creative Commons license and indicate if changes were made.

The images or other third party material in this chapter are included in the chapter's Creative Commons license, unless indicated otherwise in a credit line to the material. If material is not included in the chapter's Creative Commons license and your intended use is not permitted by statutory regulation or exceeds the permitted use, you will need to obtain permission directly from the copyright holder.

Chapter 6
An Innovative (DIW-Based) Additive Manufacturing Process

Louison Poudelet, Anna Castellví, and Laura Calvo

Abstract This chapter will describe the activity of Fenix project that consisted in developing the hardware, infrastructure and processes to make possible the re-use of the recycled metals through an Additive Manufacturing (AM) method called Direct Ink Writing (DIW). It will first explain what is DIW and why it is an interesting way to give added value to recycled materials specially metals. It will then focus on the working principles and the parts of a DIW machine and end with a conclusion of the adequacy of this technology to new circular business models for the recycling of Waste of Electric and Electronic Equipment (WEEE).

Keywords Additive manufacturing · Direct ink writing · Copper · Steel · Bimodal

6.1 Direct Ink Writing

6.1.1 DIW Technology Introduction

In a nutshell, DIW also called robocasting consists in depositing a pseudo-plastic ink composed of a solid load of particles and a binder. This paste, contained in a syringe, is deposited using a XYZ positioning system and this "green" part is then sintered in a hoven. During this process the binder is burned, and the solid particles sold together forming a solid metallic part. The solid load can be composed of metallic particles of every element and alloys, it can also be made of ceramics, glasses and last but not least biomaterials, loaded or not with living cells for tissue engineering. This section will start treating the material itself to be printed and then make a quick introduction about the design of DIW fabricated parts (Fig. 6.1).

L. Poudelet (✉) · A. Castellví · L. Calvo
Department of Research, Development and Investigation, CIM-UPC, Llorenç i Artigas, 12 08028 Barcelona, Spain
e-mail: lpoudelet@cimupc.org

© The Author(s) 2021
P. Rosa and S. Terzi (eds.), *New Business Models for the Reuse of Secondary Resources from WEEEs*, PoliMI SpringerBriefs,
https://doi.org/10.1007/978-3-030-74886-9_6

Fig. 6.1 DIW process

6.1.2 Ink Process Generation for DIW Technology

In Chap. 5 "A mobile pilot plant for the recovery of precious and critical raw materials" UNIVAQ described how the different chemical elements of the WEEEs are obtained through a hydrometallurgical process.

The powder obtained in the hydrometallurgical plant is analyzed to get information on their actual composition via EDX technique. Pre-processing steps can be put in place if the powder result to be too much oxidized (i.e. thermal treatment in Ar/H_2 atmosphere) or if the powder morphology is not suitable for the high energy ball milling step (i.e. mild powder grinding via tumbling mills).

The recycled powder is then processed with fresh raw element powders, (i.e. Fe, Ni, P) to produce an alloy suitable for sintering processes. The ratio between new and recycled materials is adjusted batch by batch according to the composition of the recycled powder. In the high energy ball milling step process, the different powders are alloyed at solid state and room temperature conditions. Once the alloyed powder is obtained it is post-processed to optimize morphology and particle size distribution.

A tumbling mill is used to increase the fraction of particles in the usable size range (i.e. particles smaller than 60 μm) and sieves are used to tailor the size distribution (i.e. bimodal or monomodal). Laser diffraction analysis assesses the final size distribution of the batch of powder that is then used to compound a feedstock for robocasting.

Once the metal powder alloy has been successfully obtained and manufactured, the next step is to formulate the appropriate composition. The formulation has been done to develop a material ink to be printed through Direct Ink Writing (DIW) process.

This process consists in the generation of an ink which its characteristics must present a pseudoplastic behavior to be printable by DIW. It is usually recommended to have a solid load by 35–60% to obtain a functional final part. During Fenix project, the development will be focused on increasing the solid load fraction in order to increase part density.

ID	Description	Input	Output	UseCase
FNX11	Fe-based powder	Raw element	Powder	Use cases One and Three
FNX12	Bimodal Fe-based powder	Raw element	Powder	Use case One - Robocasting
FNX20	Fe-based powder	Raw element	Powder	Use cases One and Three
FNX24	Bimodal Fe-based powder	Raw element	Powder	Use case One - Robocasting
FNX31	Bimodal Fe-based powder	Raw element	Powder	Use case One - Robocasting
FNX61	Monomodal Fe-based powder	LOT1	Powder	Use case One - Robocasting
FNX50	Recycled Copper	Raw element	Powder	Use case One - Robocasting
FNX51	Commercial Copper	Raw element	Powder	Use case One - Robocasting

Fig. 6.2 Powders used during the project

In the Fenix project 8 different powders have been tested, two commercial copper-based powder and six different Fe-based powders have been developed (monomodal and bimodal) (Fig. 6.2).

8 different types of inks are obtained from mixing theses powder with a binder formed of pluronic acid and a dispersing agent, with a Powder/Binder ratio of 45% in volume (82% in weight for the FNX31 powder). The following scheme presents the process of the material generation (Fig. 6.3).

The procedure to make the ink is the following:

(i)   Pluronic hydrogel with 25%w/v concentration. Selected for suitable viscoelastic properties and pseudoplastic behavior.
(ii) 45% by weight of powder (monomodal or bimodal):
(iii) Dolapix PC75 is the dispersant agent used for the composition.

As 10 ml syringes are used for the print tests, the formulation has been based on this amount. Below, the specific formulation of the inks to obtain 12 ml of mixture is presented:

- Fe-based (monomodal/bimodal) powder: 42.52 g = 5.4 ml
- Pluronic 25%: 7.26 g = 6.6 ml

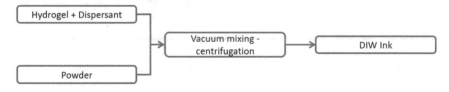

Fig. 6.3 Scheme of the process material generation

- Dolapix PC75: 0.1 g = 0.44 ml

To obtain the inks the following process is followed:

1. Blend the pluronic with the dispersant agent, dolapix PC75.
2. Let the mixture cool in a bowl with ice for 5 min.
3. Centrifugate the mixture for 2 min. The speed and power are automatically adjusted according to the weight. So, the weight of the can with the mixture and the lid must be introduced into the centrifuge in order to achieve optimum centrifugation.
4. Add the Fe-based powder to the mixture.
5. Let it cool again in a bowl with ice for 5 min.
6. Centrifugate it for 2 min again.

The equipment used for mixing the components is the centrifuge Thinky Planetary Vacuum Mixer ARV310. With the ink ready the last step consists in introducing the ink in the syringe in vacuum condition to avoid bubbles.

6.1.3 Printable DIW Parts Design Criteria

The first step to start the printing process is to know what you want to print and turn it into a 3D tangible part in stl format. It has to be designed with the right size and optimize shape according to the printing needs. So it is important to take into consideration some aspects like the limitations of the wall thickness and the minimum angles among other geometrical considerations very common in 3D printing that are described in Fig. 6.4.

6.2 Whys of DIW

DIW can be easily compared with other AM technologies and it is important to do it when choosing one.

One similar technology is FFF (fused filament fabrication) and in comparison, the big advantage is that DIW allows to print with metallic and ceramic pieces from a filled syringe without having much trouble and in an efficient process.

A good advantage of DIW compared to powder bed fusion methods like Selective Laser Sintering (SLS), Selective Laser Melting (SLM) and Electron Beam Additive Manufacturing (EBAM) is that it requires an amount of material much smaller. This is because the powder bed fusion method requires to fill all the batch of the machine meanwhile DIW just needs to fill the syringe which the minimum required is 3 ml and all will be used for the part. This makes DIW a competitive technology especially when small amounts of material are available and generates low waste. Consequently, DIW gives the best value to available material. Thus, this process is well adapted to

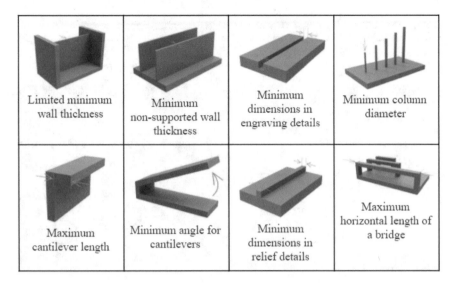

Limited minimum wall thickness	Minimum non-supported wall thickness	Minimum dimensions in engraving details	Minimum column diameter
Maximum cantilever length	Minimum angle for cantilevers	Minimum dimensions in relief details	Maximum horizontal length of a bridge

Fig. 6.4 Resume of design requirements for 3D printing

circular economy models due to the amount of material is low and the profitability depends greatly on the capacity to give the maximum added value to the extracted material.

Although it is a suitable technology it is also important to know if DIW is affordable or not. To do so, it has to be compared with another process that could manufacture more or less the same quality product. And again, the best candidate to compare DIW is Selective Laser Melting.

SLM, also known as direct metal laser melting, uses a power laser to melt and fuse powder that is commonly metal material. Even SLM has more geometry options with less design limits, both have in common the possibility of adjusting the desired infill and modifying the geometry until it is optimized. SLM uses layers from 30 to 50 μm thick and DIW layers thickness are between 0.3 and 0.5 mm. In comparison, DIW has the worst surface finish but both need post processing so at the end, this fact is not determinant.

Analyzing technological requirements, SLM has to have perfect adjustments of the optic-mechanical system like the laser spot diameter, the mode of radiation or the laser power and also, technical and environmental considerations like powder size or the airflow of the atmosphere. DIW's main requirement is ink composition to be printable, but also the hydraulic system, heating temperatures and some tip considerations like the small diameter needed. The processes are different, but it is easy to distinguish that DIW has simpler technology requirements compared with SLM. For this reason, DIW has cheaper machine costs. It is around 20.000 and 25.000 € meanwhile SLM is much more expensive as shown in Fig. 6.5.

So DIW is not only the suitable technology to develop the demonstrators, but also the most affordable in the market actually.

Range Price via Web	Approximate Price via Web	Prince Found via Website	Price obtained by E-mail
Estimated Price Resulting		Price not available	
Manufacturer	3D Printers	Information Found	Price Chosen
Renishaw	AM400	>250k$	$600,000
	AM250	$585,305	$585,305
	RenAM 500M	600K-700K$	$650,000
	Renishaw SLM 125	$300,000	$300,000
Arcam-General Electrics GE	Arcam A2X	$1,200,000	$1,200,000
	Arcam Q20Plus	$1,000,000	$1,000,000
	Arcam Q10	$1,000,000	$1,000,000
Optomec	Aerosol Jet HD(150kD)	$150,000	$150,000
	Aerosol Jet 200	$200,000	$200,000
	AerosolJet 300	$300,000	$300,000
	Lens MR-7	750-1000k$	$875,000
	AerosolJet 5X	$495,000	$495,000

Fig. 6.5 Range prices of SLS machines [1]

6.3 FENIX's DIW Machine

6.3.1 Machine Parts

The Fenix's DIW machine, designed by CIM-UPC, is composed of different modules. All of them designed according to the requirements of a DIW machine.

- Structure: The structure is made with commercial aluminum profiles subjected by angles. It has wheels with brakes to allow an easy mobility (Fig. 6.6).

Fig. 6.6 CAD detail design of the structure and its implementation

Fig. 6.7 CAD detail design and the axes implementation

- Axis: 3 axes (X, Y, and Z) are implemented. Y axis is mounted in a gantry setup (2 parallel and synchronized carriages) and XZ axis are orthogonal single carriage. Each axis possesses an end stop sensor which allows the machine to find a reference point in space from which each position of the tool head is calculated (homing process) (Fig. 6.7).
- Print head: It contains the syringe and in this design, the capacity is 10 cc, but it is possible to change it to 3 cc or 5 cc. The extrusion is volumetric with a force up to 1635 N. This force will allow to apply a pressure of 10 Bars on the ink inside a 10 cc syringe. Enough to meet the expected maximum pressure. A probe sensor is included in the extruder head and its function is to create a 3D mapping of the print base doing point mapping (Fig. 6.8).
- Construction platform: It is the base on which the part will be printed. It also has interchangeable glasses to print another construction next. It is located directly over the marble to ensure flatness (Fig. 6.9).

6.3.2 Printing Process with FENIX Machine

The process to start printing is quite similar to many other 3D printers. It all starts with an idea that needs to be tangible. After a STL with the 3D object is developed it is time to elaborate the digital file called GCode, that defines mostly instructions on where to move, how fast to move and which path to follow.

Once it is ready, the DIW printing process in the Fenix machine starts. The steps to follow are:

1. Turn on, enable the machine and plug in network wire to the laptop.
2. Start the BLTouch mapping to allow the machine to correct bed height imperfections.
3. Define Z height. It will depend every time on the size of the tip used so it is an important step to ensure the first layer is going to be well deposited.
4. Weigh the syringe, load it into the shell, place them on support by screwing and add the selected tip for the print (Fig. 6.10).

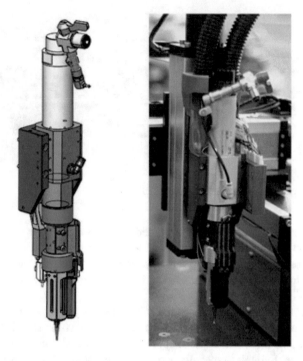

Fig. 6.8 CAD detail design and the implementation of the printer head

Fig. 6.9 The implemented construction platform

5. Check ambient temperature and if it is below 22 °C degrees turn on the bed temperature.
6. Extrude some material until it comes out correctly to ensure it is not dried at the tip and check the consistency of the ink. It has to be homogeneous and pasty (neither dry nor liquid) to be correctly printed.

Fig. 6.10 Loading syringe into the shell process

7. Finally, upload and execute the GCode of the part to print. If it is not successful, correct parameters and restart the process (Fig. 6.11).
8. Weight the syringe once finished the print to know the quantity of the material used. Verify dimensions between the printing part and the CAD.

6.3.3 First Test Validation

The next step after the realization of the device is a series of tests in order to validate that it works correctly and meets the requirements. The validation tests have been carried out only using ink loaded with recycled material and also have 2 objectives: produce a series of samples and check the influence of temperature on the ink. For this reason, it was divided into two sub-series, without temperature control and using a hot bed.

For purposes of WP6 of the Fenix project, the same sample is printed in two different directions, short and long, in order to study the effects of the internal structure on the final part (Fig. 6.12).

Fig. 6.11 First completed printing test

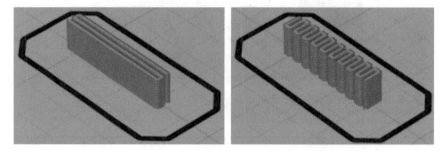

Fig. 6.12 Path planning of the sample with layers deposited in the long and short direction

After this whole process the piece is ready for the sinter process.

6.3.4 Sintering Process Parameters

After printing the green part, it is necessary to burn the binders and fuse the metallic particles together. This process is called sintering. The chart below shows the different sintering cycles used during different phases of the project. The objective of these tests are to observe the effect of the sintering and the different properties that can be given to the final material (Fig. 6.13).

The observable effects of sintering conditions are the following (Fig. 6.14).

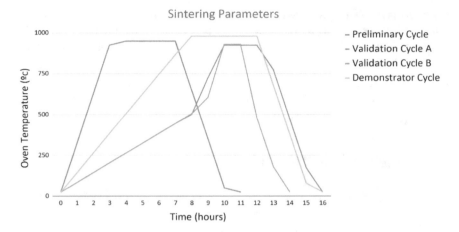

Fig. 6.13 Sintering parameters cycles

041219B08-Bimodal ink	130220B101-Monomodal ink	140220B315-Monomodal ink	290720B005-Bimodal ink
Preliminary cycle: 30 min step of 550°c and an isotherm of 2h at 925°c	**Cycle A:** An isotherm at 950°C for 4h in flux of Ar/H2 (90/10) mixture in a tubular furnace, the samples were covered in alumina sand to prevent them to lose consistency.	**Cycle B:** Consisted in an intermediate step at 530°C for 30min, and a short isotherm at 930°C for 60min	**Final demonstrators:** An isotherm at 980°C for 8h in a flux of Ar/H2 5% and a cooling ramp of 5°C/min.
Results: Really good with a very uniform and dense material.	Results: High porosity is observed.	Results: Porosity is observed.	Results: Good density in the core of the part. There is porosity in the surface.

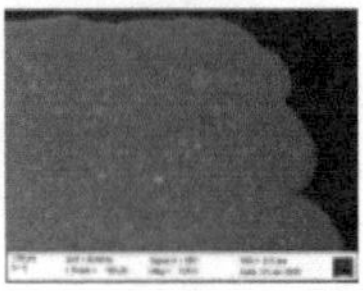

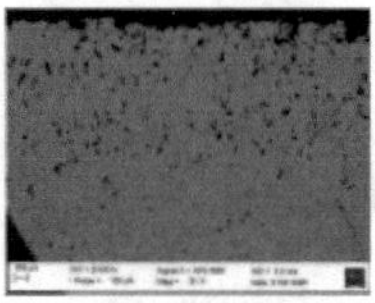

Fig. 6.14 Effects of sintering conditions

Many parameters influence the final quality of the parts but a longer cycle (especially the isotherm part) seems to be highly beneficial for the process as it can be observed between the two monomodal tests (cycle A and B) and also bimodal inks seem to have the best density results.

6.4 Technology's Viability

6.4.1 Applications in the Industry

CIM-UPC has printed two different use-case demonstrators with two different types of powders provided, FNX24 and FNX31, both Fe-Bi-modal.

The use cases demonstrators have been designed in order to prove the ability of the Fenix project to enter the industry with real applications and a circular process.

About the two demonstrators, one is a piece called Endstop and it is used in the machine itself to detect the end of the axis travel. The other one is a Handle useful for quick fixing with a thread on it. It has been machined after being printed. Below it is shown the printing results and the functional piece that would be replaced.

1. Endstop:

See Fig. 6.15.

2. Handle:

See Fig. 6.16.

Between the first tests, that were not a complex geometry, and the last ones, there have been a lot of improvements in the 3D printing process. Many printing parameters have been adjusted to the ink created. For example, the printing speed has been reduced to improve the extrusion deposition and to get more accuracy in the geometry printed. As temperature had an important role, the bed temperature was implemented and adjusted according to the ambient conditions. This made a

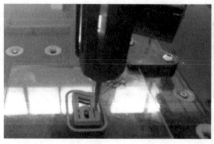

Fig. 6.15 End-stop use-case

Fig. 6.16 Handle use-case demonstration and commercial application

Fig. 6.17 Use-case demonstrator printing process

good improvement in the ink fluency and adherence during the printing. All these improvements have a positive impact and consequently functional parts have been printed (Fig. 6.17).

All data about printing conditions, characterizations and tests have been uploaded on ALBUS, that is Fenix platform to store data at all the stages of the circular business.

Half of these demonstrators have been sintered in Barcelona, same place where printed, and the others were sent to Italy to sinter. It is indicated in the ID of each part, with an A or a B respectively. This way it is possible to know if the travel and the time between the printing and the sintering, affects or not the final properties. Due to the hard travel conditions, the piece has to be very well packed with foam and other kinds of protections to ensure a good arrival, however, it has been observed that it is not guaranteed. However, this part is still in process waiting for the results.

CIM-UPC has also possible use-case demonstrators for copper inks, based on its main properties high thermal and electrical conductivity.

(i) Heat sink: it is a passive heat exchanger made by thermal conductivity materials that dissipates the heat from a mechanical or electronic device. Cooper has perfect conditions for it, however as it is less ductile than other materials like aluminum, it is difficult and more expensive to manufacture into heat sinks

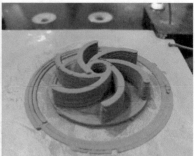

Fig. 6.18 Commercial fan-less heatsink application and copper printed demonstrator

in comparison with the aluminum ones. This is why DIW 3D printing can be the perfect technology to manufacture them and CIM-UPC has also printed one demonstrator for this use-case with a little more complex geometry. The ink used for the print has been FNX51, a commercial copper ink (Fig. 6.18).

(ii)   Electrode connector for EDM (Electrical Discharge Machining): It is the tool-electrode for a metal fabrication process based on the use of electrical discharges. This tool requires good electrical conductivity properties and complex forms, so not only copper is a good material for it but also DIW technology is a good process to get it (Fig. 6.19).

(iii)   MIG Welding Nozzle: It is used to direct the gas into the weld puddle and to protect the contact tip from molten metal. Cooper is the only candidate when the process requires higher amperage. As copper has tough machining process conditions because of its softness property, DIW technology can, again, be a good alternative to manufacture them (Fig. 6.20).

Fig. 6.19 Commercial copper tool-electrode

Fig. 6.20 Commercial copper MIG nozzle

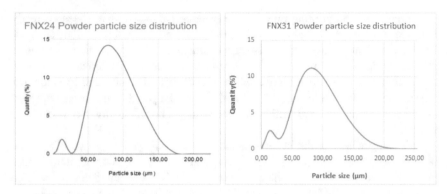

Fig. 6.21 FNX24 and FNX31 bimodal particle distribution

The last two use-case applications will be the next ones to print and test if they are viable and to test properties.

6.4.2 Applications in the Industry

Good results are linked to good part properties, with a similar density of the metal to achieve good mechanical properties. As seen before in the sintering previous section, monomodal ink samples show high porosity in the SEM (Scanning Electron microscopy) image and as a consequence, little mechanical properties. In other manufacturing processes density is increased by adding some pressure to the mold but in 3D printing this is not a possible method. Even though there are other methods, in this case, changing the distribution of the particles in the ink is the optimal one.

To do so, not only sintering parameters have been modified but also bimodal powders are developed, called FNX24 and FNX31. The main particle has a size of 82 μm and the next has around 12 μm for FNX24 and a size of 81 and 15 μm for FNX31. Consequently, the little spaces are filled with the smallest ones and density increases closer to the metal (8.13 g/cm^3). Thereby, powders have been modified to get the minimum porosities of the printed parts (Fig. 6.21).

So the good results are obtained, not only for the increased density, but also for the generation of good mechanical properties.

6.5 Conclusions

To conclude this chapter, it is worth to mention different points because the additive manufacturing process involves many different aspects such as the printing process itself, the sintering post-process, the material and the whole system based on circular economy.

Starting with the printing process, the printing parameters have been correctly determined to get good quality samples. Besides, sintering has been done successfully around Europe, overcoming shipping difficulties of the parts however due to COVID-19 this system has to be re-defined considering EU lockdown. In spite of these shipping issues, the effectiveness of the sintering is confirmed by obtaining bulk metallic parts. Moreover, the optimal size of the powders has been improved and thanks to bimodal composition, printed parts get better density and mechanical properties.

Referring to the circular system, thanks to the Fenix project, a workflow plan has been developed between different partners around Europe and with it, a circular business model has been implemented successfully. It has acquired and uploaded a complete dataset called ALBUS to track the different steps in the Fenix project to close circularities.

And last but not least, DIW technology has proven to be the best suitable additive manufacturing technology to recycle WEEE by developing the demonstrators, not only with Fe-based ink but also with copper-ink, with real applications.

Reference

1. Peroncini, S. (2018/2019). *European University market for 3D printing and business plan of an additive manufacturing laboratory* (Online).

Open Access This chapter is licensed under the terms of the Creative Commons Attribution 4.0 International License (http://creativecommons.org/licenses/by/4.0/), which permits use, sharing, adaptation, distribution and reproduction in any medium or format, as long as you give appropriate credit to the original author(s) and the source, provide a link to the Creative Commons license and indicate if changes were made.

The images or other third party material in this chapter are included in the chapter's Creative Commons license, unless indicated otherwise in a credit line to the material. If material is not included in the chapter's Creative Commons license and your intended use is not permitted by statutory regulation or exceeds the permitted use, you will need to obtain permission directly from the copyright holder.

Chapter 7
The Life Cycle Performance Assessment (LCPA) Methodology

Reinhard Ahlers

Abstract The FENIX project has started to develop future business models for the efficient recovery of secondary resources. It would not be enough just to improve business models based on traditional linear approaches. Rather, new approaches must be developed with a particular focus on environmental and climate changes. Electronic scrap is no longer scrap, but must be seen as valuable material. Using the mobile phone as an example, FENIX has developed technologies to get recyclable materials out of scrapped mobile phones and to process them into new materials and final products. The developed technological approaches are not limited to mobile phones, but can be used for all types of electronic waste. FENIX has only focused on the logistic chain from the dismantling of the cell phones to the manufacturing of new materials and products (recycling chain). This, of course, involves a lot of effort in dismantling the e-waste, as the recycling process was not yet considered when developing the products currently on the market. Such eco-design approaches would certainly reduce the disassembly effort in the future. FENIX business models should not only be based on economic success but also consider ecological effects at the same time. Therefore, an accompanying Life Cycle Performance Assessment (LCPA) has been carried out to prove the advantages of the developed business models. From the interim assessment, recommendations for further technical development directions were repeatedly given to achieve the best possible economic and ecological solutions.

Keywords Life cycle performance analysis · BAL.LCPA · Circular business models · Key Performance Indicators (KPIs) · Net Present Value (NPV) · Greenhouse Warming Potential (GWP) · Cumulative Energy Demand (CED) · External costs

R. Ahlers (✉)
BALance Technology Consulting GmbH, Contrescarpe 33, 28203 Bremen, Germany
e-mail: reinhard.ahlers@bal.eu

© The Author(s) 2021
P. Rosa and S. Terzi (eds.), *New Business Models for the Reuse of Secondary Resources from WEEEs*, PoliMI SpringerBriefs,
https://doi.org/10.1007/978-3-030-74886-9_7

7.1 Sustainable Business Models

The future of companies, the environment, and society depends on sustainable business models. David and Martin [1] outlined a corporate management strategy to create new modes of differentiation, embedding societal value into products and services, reshaping business models for sustainability and define new measures of performance.

The treatment of e-waste will get more important for the preservation of natural resources. The FENIX business models are focusing on:

- cooperation in recycling and production beyond company boundaries,
- definition of optimal logistical processes and the,
- utilisation of recycled materials from e.g., electronic items for new products.

In general sustainability refers to four distinct areas: economic, environmental, social and human as defined by the RMIT University [2].

Economic sustainability: Economic sustainability aims to maintain the capital intact and to improve the standard of living. In the context of business, it refers to the efficient use of assets to maintain company profitability over time. But the approach that continuous growth is good even when it harms the ecological and human environment is becoming less important. New economics approaches include also natural capital (ecological systems) and social capital (relationships amongst people).

Environmental sustainability: Environmental sustainability aims to improve human welfare through the protection of natural resources (e.g. land, air, water, minerals etc.). The consideration of environmentally sustainability lowers the risk of compromising the needs of future generations. It has to be considered how business can achieve positive economic outcomes without doing any harm, in the short or long-term, to the environment.

Social sustainability: Social sustainability aims to preserve social capital by investing and creating services that constitute the framework of our society. This requires a larger view of the world in relation to communities, cultures and globalisation. Social sustainability focuses on maintaining and improving social qualities like cohesion, reciprocity, social equality, honesty, and the importance of relationships amongst people.

Human sustainability: Human sustainability aims to maintain and improve the human capital in society. Investments in health and education systems, access to services, nutrition, knowledge, and skills are examples for human sustainability. In the context of business, an organisation will view itself as a member of society and promote business values that respect human capital. Human sustainability focuses on the importance of anyone directly or indirectly involved in the making products or offering services.

All four sustainability areas have been considered to create new products from e-waste in FENIX. Nevertheless, the LCPA assessment activities were mainly focused on the economic and environmental aspects [3].

7.2   Electrical and Electronic Waste Market

Waste of Electrical and Electronic Equipment (WEEE) is a complex mixture of materials and components which can partly be recycled and reused. Another part of the waste contains hazardous materials which can cause major environmental and health problems if not managed in a proper way. WEEE includes e.g. computers, TV-sets, fridges, washing machines, desktop PCs, notebooks and mobile phones. The waste of electrical and electronic equipment is one the fastest growing waste streams in the EU, and it is expected that it will grow to more than 12 million tons by 2020 [Source: https://ec.europa.eu/environment/waste/weee/index_en.htm].

EUROSTAT (EUROpean STATistical Office) estimates that the second and third largest categories for WEEE (Waste Electrical and Electronic Equipment) collection in the EU comprises around 555 thousand tonnes of consumer equipment and photo-voltaic panels (14.8%) followed by IT and telecommunications equipment (14.6%) with 547 thousand tonnes.

The production of modern electronics requires the use of scarce and expensive resources (e.g. around 10% of total gold worldwide is used for electronic equipment production). To improve the environmental management of WEEE and to contribute to a circular economy and enhance resource efficiency the improvement of collection, treatment and recycling of electronics at the end of their life is essential.

Environmental risks may take place in the cases where e-waste is not handled properly within the recycling and pre-treatment processes. With proper technologies, 100% of the materials in a mobile phone can be recovered and nothing needs to be wasted. In the first approach the FENIX project focuses on the valuable materials of the mobile phones (gold, silver, copper, etc.). But it can easily applied to other kinds of WEEE.

The main challenge of the old mobile phone collection is to get people to return their old products for recycling when they no longer need them. One inhibiting factor for recycling of mobile phones is the willingness to keep a spare product. The most important factors enhancing the recycling behavior are convenience and awareness on where and how to recycle.

Mobile phones are just one example in a high varity of electronic products. Never-theless it is one of the products with the most valuable materials inside (see Table 7.1). The following table shows the average material content in different product catagories. The FENIX pilot oprations have shown that it can differer very much from batch to batch. Therefore the values can only be used as guidelines. The large amount of material to be expected in a mobile phone was a reason to choose these items. During the assessment of the different FENIX processes it has been shown

Table. 7.1 Characterization of metals embedded in specific WEEE

Average material content in Prm (Percentage of recycled materials)	Refrigerator	Wasching machine	Air conditioner	Desktop PC	Notebook	Mobile phone	CRT TV	Stereo system	Digital camera
Iron (Fe)	2.1	9.5	2.0	1.3	3.7	*1.8*	3.4	1.2	3.0
Copper (Cu)	17.0	7.0	7.5	20.0	19.0	**33.0**	7.2	15.0	27.0
Silver (Ag)	0.0	0.0	0.0	0.1	0.1	**0.4**	0.0	0.0	0.3
Gold (Au)	0.0	0.0	0.0	0.0	0.1	**0.2**	0.0	0.0	0.1
Aluminium (Al)	1.6	0.1	0.7	1.8	1.8	*1.5*	6.2	2.9	2.4
Barium (Ba)	0.0	0.0	0.0	0.2	0.6	*1.9*	0.2	0.1	1.6
Chromium (Cr)	0.0	0.0	0.0	0.0	0.1	*0.1*	0.0	0.0	0.3
Lead (Pb)	2.1	0.2	0.6	2.3	1.0	*1.3*	1.4	1.9	1.7
Antimony (Sb)	0.3	0.0	0.0	0.2	0.1	*0.1*	0.3	0.0	0.2
Tin (Sn)	8.3	0.9	1.9	1.8	1.6	*3.5*	1.8	2.2	3.9
Zinc (Zn)	1.7	0.2	0.5	0.3	1.6	*0.5*	5.3	1.4	0.9

Source Cucchiella et al. [4]

Remark: 0.2% of Au in 1 ton of PCBs means 200 g of Au in 1 ton of PCBs

that the material composition of the e-waste has an high impact on the economic efficiency of the examined processes.

7.3 Life Cycle Performance Assessment (LCPA) for FENIX

The ecological awareness of customers is increasing in Europe. The FENIX project has started to improve the recycling processes and to make better use of electronic waste using cell phones as an example. The project has focused on the optimization of the recycling processes and process chains starting from disassembly up to the production of recycled materials and products.

Different approaches and technologies for disassembly, recycling, and up-scaling of recycled material have been tested. These FENIX processes are interconnected and form three supply chains with the aim of creating three different products (jewelry, filament for additive manufacturing and ink for additive manufacturing). All three implemented supply chains started with the disassembly processes of mobile phones and followed by the recycling process. While the recycling process delivers the gold material, extracted from the e-waste directly to the jewelry production, the other extracted materials (mainly copper) are delivered to an up-scaling process. Within this process the copper is prepared to produce ink and advanced filaments for additive manufacturing.

To compare the different approaches and technologies developed by FENIX and to verify the economic viability as well as the ecological impact an LCPA (Life Cycle Performance Assessment) has been performed. The Life Cycle Performance Assessment (LCPA) includes the Life Cycle Assessment (LCA) with the focus on the ecological impact and the Life Cycle Cost analysis (LCC) considering the economic calculations [5].

The LCA is defined as compilation and evaluation of in- and outputs (e.g., use of natural resources, emissions to air, water, and waste) and the potential environmental impacts throughout its life cycle. The LCC approach has been used to analyse the economic perspective by applying the Net Present Value (NPV). The NPV is calculated by a dynamic procedure and considers the current value at each time which means that earlier revenues are valued higher than later ones. To prove the profitably, the calculation of the net present value is essential.

The most important KPIs (Key Performance Indicators) have been defined in cooperation with the different FENIX process owners (disassembly, recycling and up-cycling) (see Table 7.2). LCPA results base on a combination of various KPIs including life cycle costs, Global Warming Potential and the cumulative energy demand.

For the FENIX assessment the following Key Performance Indicators (KPIs) have been selected from a larger set of parameters.

The assessment bases on complex mathematical models. To carry out the LCA and LCC assessment in parallel the commercial LCPA tool from BALance (BAL.LCPA)

Table 7.2 KPIs selected for the FENIX assessment

Focus	KPIs	Description
LCA	Amount	Amount of materials used in in the recycling process
	Raw material to process	Indicate the materials involved like e.g. metals, minerals, plastics, textile, organic and inorganic intermediate products, paints, etc
	Electricity	Specify the Grid Mix indicating the country, or the specific mix known (e.g. 40% nuclear, 60% hydroelectric)
	Water consumption	Indicate water consumption for the production
	Generated waste	Define waste typology (e.g. plastic, inert, hazardous, metals, wastewater, liquid, emission)
	GWP	Greenhouse Warming Potential → Climate Change
	CED	Cumulative energy demand → Depletion of energy resources (distinguished between fossil and renewable energy)
	AFP	Aerosol formation potential → Damage to human health due to particular matters
	AP	Acidification potential
	EP	Eutrophication potential
LCC	NPV	Net-present value—some future value of the money when it has been invested
	External costs	E.g., costs for environmental damages
	Payback time	Period required to recoup the money expended in an investment
	Amortisation	Spreading the cost of an intangible asset over a specific period

has be applied to support a comprehensive decision-making process for process alternatives already in the early development phase of the project.

Therefore, assessment models for the different FENIX areas have been defined and were implemented in the BAL.LCPA tool. The models include the descriptions of the operational processes but also reference processes to be able to compare different approaches. The models were supplied with estimated values and later with actual measured values generated by the installed pilots to ensure realistic statements. For the FENIX pilots a screening LCA has been applied to focus on the most important environmental challenges.

The BAL.LCPA software tool allows the quick adaptation of the models due to pilot implementation changes and the definition of additional assessment parameters. The different assessment results are visualized and stored in the database for further use. The challenge of the assessment is to analyse each process individually to identify improvement potentials but also to optimize the entire supply chain.

7.4   Assessment of FENIX Implementations

The Life Cycle Cost analysis has been carried out for each process and the interconnections of the processes. The assessment has been divided into the processes (disassembly, recycling, and upscaling) and the product related use cases (metal powder and robocasting, jewelry production and advanced filament production). After the optimization, a comprehensive assessment was carried out. The ecological analysis (LCA) has been focused for the whole process chain starting from the disassembly process up to the material recycling/up-scaling process.

Measurements at the pilot installation have been used for the assessment as well as market figures were relevant. Because of the amount of assessment parameters only the most relevant results are summarized within this chapter.

The assessment starts with the disassembly process. FENIX is not focusing on the e-waste collection process while the improvement potentials under FENIX main emphasis is very low compared to the conventional processes of today (Fig. 7.1).

The focus of the disassembly process is to dismantle the mobile phone scrap in an environmentally friendly and cost-effective manner. Poor manual disassembly processes have been assessed as well as COBOT (COllaborative RoBOT) supported manual processes. The dismantled parts should be optimally prepared for the following FENIX recycling processes. The recycling process requires PCBs with rich materials. Batteries and cooling elements do not contribute to the extraction of valuable material. Capacitors even worsen the FENIX recycling processes. The disassembly assessment results can be briefly summarized in the following points:

- Poor manual driven disassembly processes are beneficial after a short time (months) while the duration depends mainly on the salary rate of the personal.
- Disassembly processes based on COBOT operations (manual plus robot) are too expensive under all circumstances and become never beneficial. The reasons are the high process time per mobile phone for the COBOT and the hardware investment costs.

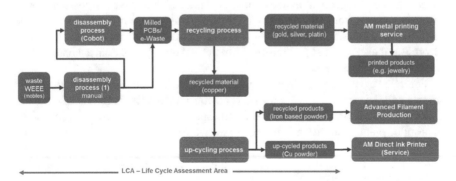

Fig. 7.1 Life Cycle Assessment area

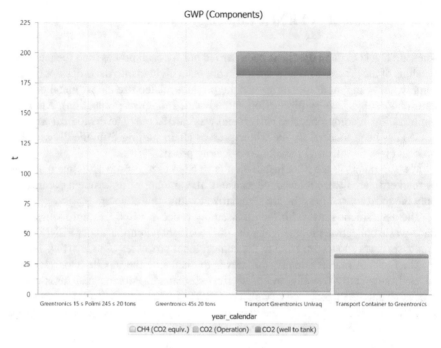

Fig. 7.2 CO_2 emissions

- Transportation costs have been calculated based on the manual disassembly process. Their influence on the NPV is very low over the evaluated period (15 years) so that it can be neglected.
- Transportation has an important influence on the GWP (Green Warming Potential) as shown as part of the LCA analysis (Fig. 7.2).

One challenge of the FENIX project was the development of a mobile recycling plant (recycling reactor in a size of one container). This makes it possible not to bring the e-waste from the disassemble service provider to the recycler, but rather the recycling process to the disassembly provider. This approach reduces the logistical effort. But it also assumes that the disassembler has enough material available to use the system for a certain period.

For the evaluation of the logistic processes the real distances between the disassembly and the recycling location have been considered as basis for the assessment. The GWP calculation bases on monthly transport of e-waste to the hydrometallurgical pilot plant. The distance between the two processes is about 1.700 km and during the transport more than 200 t GWP are produced during the 15 years. The alternative is the transport of the plant to the e-waste once a year and operated the system at the location of the collector.

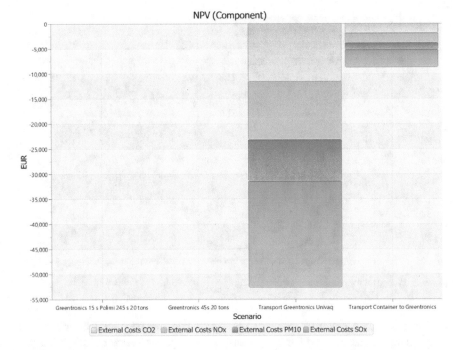

Fig. 7.3 External costs

- There is no noticeable cost difference for the operators of the processes, but the assessment shows a big difference in the external costs. These are costs that are paid by the society (e.g. health consequences of pollution).
- External costs will only become important if the saving of CO_2 is are rewarded and will affect profitability of business processes.

The following figure shows the results of the external cost assessment based on transportation (Fig. 7.3).

The main goal of the recycling process is to remove as much valuable material as possible from the e-waste prepared by the disassembly process. The hydrometallurgical pilot plant developed within the FENIX project should assure an environmentally friendly and cost-effective recycling process. The pilot installation has been focused mainly on the generation of gold, silver and copper as basis for the assessment. But other materials could also be extracted with the same unit in the future.

The recycling process assessment results can be briefly summarized in the following points:

- The semi-automated material recovery plant operated in two shifts will not become beneficial (yellow curve in the following figure). The semi-automated process was installed in the first development step of FENIX, but it became clear very quickly that a higher automation degree for the plant is required.

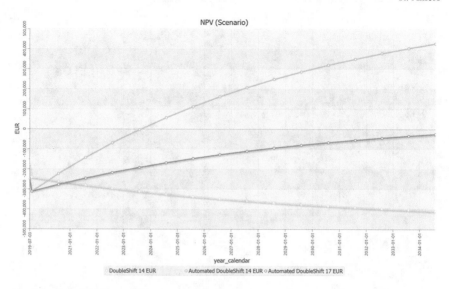

Fig. 7.4 Overall NPV scenarios

- The automated material recovery plant will become beneficial after 5 years considering the actual PCB purchasing market prices (price of the incoming e-waste).
- The sensitivity analysis for automated material recovery plant has shown that the increase of the PCB purchasing price of 10% extends the duration to 14 years before becoming beneficial. This shows the high impact of the e-waste PCB price on the economic efficiency of the process.
- Richer e-waste materials would shorten the time significantly.
- The assessment of the recycling process bases on a yearly process volume of 20 t/year (PCB waste). This volume can be achieved with a container-based reactor. Turning away from the container approach would lead to a higher process volume and thus to increase the profitability (Fig. 7.4).

The upcycling process is a preliminary stage to refine copper from the FENIX recycling processes to produce copper-based powder. This metal powder is the basis for the ink production (FENIX use case 1: Direct Ink Writers) and the production of advanced metal-based filaments (FENIX use case 3). Additionally, the metal powder should be directly sold to the market for e.g., laser metal deposition and sintering.

High energy ball milling is the central process to produce copper-based powder (pure or mixed) for different applications. The recycled powder is processed with fresh raw element powders, (i.e. Fe, Ni, P) to produce an alloy suitable for sintering processes, the ratio between pristine and recycled materials is adjusted batch by batch according with the composition of the recycled powder.

The upscaling process assessment results can be briefly summarized in the following points:

- Profitability of the upcycling process depends very much on the output quantities. Official market price for copper and additional materials have been used for the assessment.
- A minimal material output of 2 t per year and a much lower personal effort (industrial production) will assure a payback time after 8.5 years. So far, the production has only assessed on a laboratory level with a small amount of material and a high personal effort).

Looking only at the processes (disassembly, recycling and upscaling) the assessment results show that the profits are associated with different risks. This includes the market prices for e-waste and raw materials (to sell) as well as the production capacity. The processes are only profitable from a certain amount of material that must be sold on the market. To reduce the risk FENIX has also focused on products produced from recycled FENIX materials within the three use cases.

The metal powder and robocasting use case consists of the development of DIW (Direct Ink Writing) printers for high precise printings (robocasting). The DIW is a 3D-Printing technology which a paste-like filament is extruded from a small nozzle. The nozzle moves across the printing table. The new DIW developed by FENIX works with a pressure of 198 bars (state of the art DIW work with 6 bars) and can produce a higher surface quality and a more precise printing (Fig. 7.5).

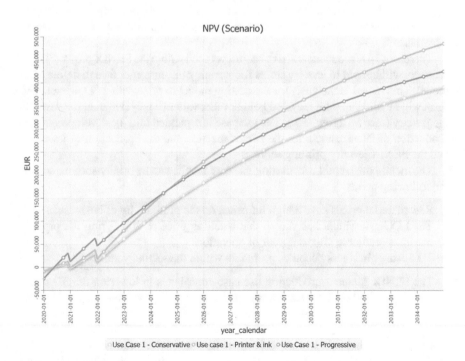

Fig. 7.5 Use case 1 NPV scenarios

Beside the development and the marketing of the printer, FENIX has developed and produced ink for the DIW printers from the materials of the FENIX upscaling process. This special ink has been optimized to enable lower sintering temperatures for the printed products. This means that smaller sintering furnaces with lower energy requirements can be used.

The combination of a high-quality printer and ink made from recycled material encounters a gap in the market that will generate greater demand in the future and promises higher margins.

The metal powder and robocasting use case results can be briefly summarized in the following points:

- The sale of the printers and the associated special ink generated from recycled material promise to be a success story. However, it must be considered that this is a new product for which only limited market figures are available.
- By the combined marketing of ink and printer the income is much higher than for recycled material. Different scenarios have been calculated (see figure). For the conservative scenario with an amount of 3 sold printers plus ink the use case becomes beneficial after one year.

The jewelry production use case has been started to use the valuable materials of the FENIX recycling process (gold, silver, etc.) to produce personalized jewelries. It is expected to generate higher margins (compared to simple recycled materials) by creating sustainable products through personalization and the use of recycled materials.

Therefore 3-D face scanners have been developed within the FENIX project. These scanners will be sold to jewelry stores for scanning the customer face to define a 3D model. This model is the basis for the casting model to print with a 3D printer. The form will be filled up with recycled gold or other valuable recycled material to make the jewelry (face on a ring, etc.). The use case is separated into the development and production of face scanners for the personalization and the production of jewelries and the FENIX jewelry printing service.

The metal powder and robocasting use case results can be briefly summarized in the following points:

- Raw material prices have a high influence on the profitability of business model.
- The LCC assessment has shown that a selling price of 200 €/ring the product becomes profitable after one year of operation.
- 3D Scanner business becomes profitable within the second year.

The FENIX filament production use case contributes to lowering the 3D metal printing costs. Today, 3D metal printing cost are very high because of the filament costs but also because of expensive industrial hardware. The FENIX filament enables 3D metal printing on conventional printers and therefore lowering the costs for 3D metal printing substantially. Low-cost metal filaments which can be used with relatively low-cost hardware and which is reliably extrusion is the competitive advantage of the FENIX filament produced from recycled materials.

The filament production use case results can be briefly summarized in the following points:

- Similar to the upscaling process the profitability of the metal filament production process depends very much on output quantities.
- The payback time for the yearly production volume of 1.8 t can be realised after 4.5 years.
- To reach the payback for lower quantities (e.g., 900 kg per year) solutions have to be found to reduce the equipment investment costs.

The most profitable use cases are the ones were the recycled materials can be distributed on the market combined with related products (e.g. jewelry, new generation of printers, etc.). A joint venture of the FENIX process owners would reduce the generated surpluses of each process but would also lower the business risk for the previous processes (disassembly, recycling, and upscaling). In summary it would lead to a beneficial recycling process chain with one overall margin and the chance of a comprehensive control over all chain elements.

7.5 LCA Assessment of the FENIX Processes and Use Cases

The environmental assessment is carried out across all FENIX processes and use cases. The highest impact has the recycling process based on the hydrometallurgical pilot plant. The pilot uses several chemical substances, energy, and water in a higher amount than the other processes and therefore dominates the LCA calculation. The following parameters have been selected for the FENIX assessment.

- Greenhouse warming Potential (GWP)
- Cumulative energy demand (CED)
- Aerosol formation potential) (AFP)
- Acidification potential (AP)
- Eutrophication potential (EP).

The benchmark for the LCA are the conventional mining processes. If not working with recycled materials, the raw materials offered by the mining industry would be the alternative.

The most important parameter within the ecological assessment is the GWP (Global Warming Potential) which describes the contribution of the recycling processes to the global warming of the earth (Fig. 7.6).

The new FENIX recycling process include the disassembly part (orange) and the recycling and upcycling processes (green). The FENIX recycling and upcycling processes are 20% better than the conventional mining process in respect of the GWP. The AFP (Aerosol Formation Potential) assess the ability of VOCs (Volatile organic compounds). VOCs are easily become gases or vapors and contribute to the formation of tropospheric ozone and smog (Fig. 7.7).

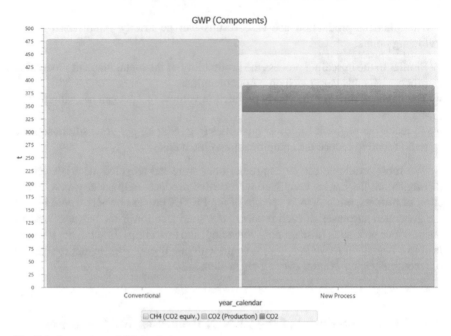

Fig. 7.6 Comparison of CO_2 emissions

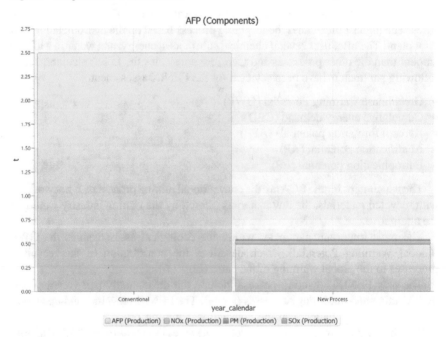

Fig. 7.7 Comparison of other emissions (1/2)

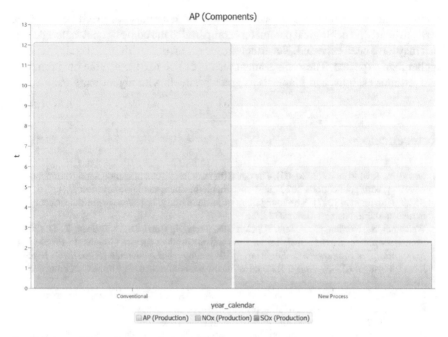

Fig. 7.8 Comparison of other emissions (2/2)

The AFP shows the greater difference between conventional and the new processes. The conventional mining processes include also NOx, PM (Particle Matters) and SOx, but in comparison to AFP they are no longer shown in the following figure. It has to be noted that NOx, PM and SOx together reach a value of 55 kg over 15 years. The AFP of the FENIX recycling process only accounts for 20% of the conventional mining process. The AP (Acidification Potential) increases leaching behavior of heavy metals in soil and has a negative impact on animals and plants (Fig. 7.8).

7.6   Conclusions

The FENIX recycling process contributes 80% less to the Acidification Potential than the conventional mining process and therefore has a significantly lower impact on the health of animals and plants. The EP (Eutrophication Potential) describes the degree of the ecosystem pollution. It shows in which the over-fertilization of water and soil has turned into an increased growth of biomass. Conventional processes generate no EP, while the FENIX recycling processes have a very small share (5.6 kg). The Eutrophication Potential is the only ecological parameters were the FENIX processes are worse than the conventional processes but on a very low level. The LCA in FENIX compares the use of recycled materials with the conventional raw materials from mining. This assessment has shown that recycled materials are much better (up to 80%) in nearly all ecological parameters compared to the conventional material. This fact may have been expected. Nevertheless, the extent of the difference is significantly higher than expected. These assessment results can be used to sell the final products like jewelleries, inks, and filaments as green products with higher margins.

References

1. David, Y., & Martin, R. (2020–03). The quest for sustainable business model innovation. https://www.bcg.com/publications/2020/quest-sustainable-business-model-innovation
2. RMIT University. (2017). The four pillars of sustainability. https://www.futurelearn.com/courses/sustainable-business/0/steps/78337
3. Wellsandt, S., Norden, C., Ahlers, R., Corti, D., Terzi, S., Cerri, D., & Thoben, K.-D. (2017). Model-supported lifecycle analysis: an approach for product-service systems. In *Proceedings of International Conference on Engineering, Technology and Innovation (ICE/ITMC)*, Madeira Island, Portugal (27–29 June 2017). https://www.ice-conference.org/Home/Conference-2017.aspx
4. Cucchiella, R., D'Adamo, I., Koh, S. C. L., & Rosa, P. (2016). A profitability assessment of European recycling processes treating printed circuit boards from waste electrical and electronic equipment. *Renewable and Sustainable Energy Reviews*, 749–769.
5. Ahlers, R., Fontana, A., Petrucciani, M., Cassina, J., Corti, D., & Norden, C. (2017). Synchronised monitoring of sustainability and life cycle costs with a modular maritime IT-platform. In RINA (Ed.), *Proceedings of the 18th International Conference on Computer Applications in Shipbuilding*, Singapur, Singapur, 26–28 September 2017 (Vol. II, pp. 91–101). ISBN 978-1-909024-67-0.

Open Access This chapter is licensed under the terms of the Creative Commons Attribution 4.0 International License (http://creativecommons.org/licenses/by/4.0/), which permits use, sharing, adaptation, distribution and reproduction in any medium or format, as long as you give appropriate credit to the original author(s) and the source, provide a link to the Creative Commons license and indicate if changes were made.

The images or other third party material in this chapter are included in the chapter's Creative Commons license, unless indicated otherwise in a credit line to the material. If material is not included in the chapter's Creative Commons license and your intended use is not permitted by statutory regulation or exceeds the permitted use, you will need to obtain permission directly from the copyright holder.

Chapter 8
A Decision-Support System for the Digitization of Circular Supply Chains

Dimitris Ntalaperas, Iosif Angelidis, Giorgos Vafeiadis, and Danai Vergeti

Abstract As it has been already explained, it is very important for circular economies to minimize the wasted resources, as well as maximize the utilization value of the existing ones. To that end, experts can evaluate the materials and give an accurate estimation for both aspects. In that case, one might wonder, why is a decision support system employing machine learning necessary? While a fully automated machine learning model rarely surpasses a human's ability in such tasks, there are several advantages in employing one. For starters, human experts will be more expensive to employ, rather than use an algorithm. One could claim that research towards developing an efficient and fully automated decision support system would end up costing more than employing actual human experts. In this instance, it is paramount to think long-term. Investing in this kind of research will create systems which are reusable, extensible, and scalable. This aspect alone more than remedies the initial costs. It is also important to observe that, if the number of wastes to be processed is more than the human experts can process in a timely fashion, they will not be able to provide their services, even if employment costs were not a concern. On the contrary, a machine learning model is perfectly capable of scaling to humongous amounts of data, conducting fast data processing and decision making. For power plants with particularly fast processing needs, an automated decision support system is an important asset. Moreover, a decision support system can predict the future based on past observations. While not always entirely spot on, it can give a future estimation about aspects such as energy required, amounts of wastes produced etc. in the future. Therefore, processing plants can plan of time and adapt to specific needs. A human expert can provide this as well to some degree, but on a much smaller scale. Especially in time series forecasting, it is interesting to note that, even if a decision support model does not predict exact values, it is highly likely to predict trends of the value increasing or decreasing in certain ranges. In the next sections, we are going to describe the four machine learning models that were developed and which compose the Decision Support System of FENIX. Section 8.1 describes how we predict the quality of the extracted materials based on features such as temperature, extruder speed, etc. Section 8.2 describes the process of extracting heuristic rules based on

D. Ntalaperas (✉) · I. Angelidis · G. Vafeiadis · D. Vergeti
SingularLogic, Achaias 3 & Trizinias st., 14564 Kifissia, Greece

© The Author(s) 2021

P. Rosa and S. Terzi (eds.), *New Business Models for the Reuse of Secondary Resources from WEEEs*, PoliMI SpringerBriefs,
https://doi.org/10.1007/978-3-030-74886-9_8

existing results. Section 8.3 describes how FENIX provides time-series forecasting to predict the future of a variable based on past observations. Finally, Sect. 8.4 describes the process of classifying materials based on images.

8.1 Extracted Materials Quality Prediction

The first model of the proposed decision support system predicts the quality of the produced material based on input features. More specifically, a Logistic Regression model is used. This model is well-documented [1] and considered a standard in deep learning. Logistic regression is used in various fields, including machine learning, most medical fields, and social sciences, while it also provides invaluable predictions in market applications. Before explaining the finer details of FENIX's model, we will explain the fundamentals of logistic regression in general. The goal of logistic regression is to find the best fitting but biologically reasonable model to describe the relationship between the binary characteristic of interest (dependent variable = response or outcome variable) and a set of independent (predictor or explanatory) variables. Logistic regression generates the coefficients, its standard errors as well as the significance levels of a formula to predict a logit transformation of the probability of presence of the characteristic of interest:

$$logit(p) = b_0 + b_1 X_1 + b_2 X_2 + b_3 X_3 + \cdots + b_k X_k$$

It originates from statistics and, while the most basic type of LR is the binary LR, which classifies inputs into one of two categories as explained above, its generalization classifies inputs into arbitrarily many categories. It is important to note, however, that machine learning does computations in terms of numerical matrices which are composed of features and weights. Since all features must be numbers, any input feature which is not a number must be properly processed to remedy this. A typical approach to this is one-hot categorical encoding. All non-numerical features in FENIX such as the name of the material have a finite set of possible values, enabling the use of one-hot categorical encoding. What this encoding does is it maps each string value into a vector of zeros, as many as the possible values for that feature, while one of them is one for the position that represents the original string. For example, let us assume that feature "f" has possible values "A", "B", "C", then we would map those to [1, 0, 0], [0, 1, 0] and [0, 0, 1], respectively. When encountering an input of "B", we would immediately convert it to [0, 1, 0]. While one-hot encoding solves the issue, it creates another one. Since other features are already numerical, we need to somehow "merge" the dimensions of the features for input X. For example, let's assume that X has the features "mean temperature" (value 34), "process energy" (value 25) and "f" (value "B"). Then, 34, 25 and [0, 1, 0] need to be fed into the model, but their dimensions do not match. This is solved by assuming each feature being a $1 \times N$ vector and then concatenating these vectors to form the final input X. In this instance, X would be [34, 25, 0, 1, 0]. Another important aspect of the model to discuss involves its activation function. The activation function defines the output of that node given an input or set of inputs. For a binary LR, the activation function would be a sigmoid, because it outputs a value in [0, 1], expressing a probability.

$$\varphi(u_i) = (1 + e^{-u_i})^{-1}$$

While this works well when a problem has only 2 possible classes, we need a different way to achieve the same result for arbitrarily many classes. The solution lies in using the softmax activation function.

$$f_i(\vec{x}) = \frac{e^{x_i}}{\sum_{j=1}^{J} e^{x_j}}, \quad j = 1, \ldots, J$$

The softmax activation function has a form that forces the resulting numbers to be in [0, 1] as well, but with one additional property: their sum is always 1, so all outputs express a probability distribution for the initial input. For example, if we have three possible classes, C1, C2, C3 and an input X, the output would be something like [0.3, 0.3, 0.4]. All values are in [0, 1] and their sum is 1. For FENIX's purposes, we try to predict the Satisfactory status of a material based on features such as material input, process mean temperature, extruder speed (mm/min). Each material is classified into one of three categories "Yes", "No", "Printable". Features which are not initially numerical are converted into one-hot categorical encodings and then all features for each input are concatenated to form the input vector for the model, following the procedure that was explained above. After the models' computations, we obtain an output of the format [p1, p2, p3], where p1 indicates the estimated probability for "Yes", p2 for "No" and p3 for "Printable". Obviously, the highest among the values is declared the model's prediction of class for the specified input. Before the model can be used, however, it needs to be trained. To that end, we conducted supervised learning. In general, machine learning uses two general learning strategies during training, depending on the task at hand: supervised or unsupervised learning. Unsupervised learning is a type of machine learning that looks for previously undetected patterns in a data set with no pre-existing labels and with a minimum of human supervision. In contrast to supervised learning that usually makes use of human-labelled data, unsupervised learning allows for modelling of probability densities over inputs. Supervised learning is the task of learning a function that maps an input to an output based on example input–output pairs. For FENIX, we already know that each material will have one of three "Satisfactory" status, we only need to learn to predict it. To that end, we labelled a generous amount of input with the correct satisfactory status. These samples were further split into test, training, and validation datasets. The split is necessary, because we want the model to train under part of the data and not overfit, which is why we need the test-train split. However, we also want to make sure its performance is sufficient even on samples it sees for the first time (which will be the case in deployment as well), as well as ensure the model does not create correlations of any kind between test and training data. This

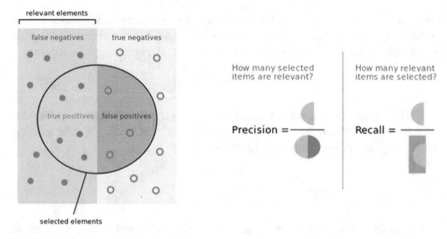

Fig. 8.1 Precision and recall

justifies the further split of the test data into test and validation. In order to increase the credibility of the results, we used state-of-the-art metrics to evaluate our model, namely precision, recall and F1-score. In pattern recognition, information retrieval and classification (machine learning), precision (also called positive predictive value) is the fraction of relevant instances among the retrieved instances, while recall (also known as sensitivity) is the fraction of the total amount of relevant instances that were retrieved. Both precision and recall are therefore based on an understanding and measure of relevance (Fig. 8.1).

Intuitively, precision shows the ratio of the outputs we predicted correctly compared to the sum of correct predictions and false positives (values that were classified as being correct by mistake). Recall shows the ratio of the outputs we predicted correctly compared to the total amount of correct predictions and false negatives (values that should be classified as the specific class but did not by mistake).

Precision formula

$$precision = \frac{true\ positives}{true\ positives + false\ positives} \tag{8.1.1}$$

Recall formula

$$recall = \frac{true\ positives}{true\ positives + false\ negatives} \tag{8.1.2}$$

Finally, F1-score belongs to the family of F metrics, which indicate a weighted mean calculation of precision and recall, with the goal of providing a final "overall score" for a model.

$$F_\beta = (1 + \beta) \cdot \frac{precision \cdot recall}{\beta^2 \cdot precision + recall} (in\ precision - recall\ terms)$$

$$F_\beta = (1+\beta) \cdot \frac{(1+\beta)^2 \cdot TP}{(1+\beta)^2 \cdot TP + \beta^2 \cdot FN + FP}(in\ TP - FP - FN\ terms)$$

When $\beta = 1$ we get the F1-score, which is the harmonic mean of precision and recall:

$$F_1 = 2 \cdot \frac{precision \cdot recall}{precision + recall} = \frac{2 \cdot TP}{2 \cdot TP + FN + FP}$$

Finally, to make the evaluation even more thorough, we conducted k-fold cross validation. This is a powerful technique when training a model under a dataset because it tries to reduce overfitting as much as possible. Simply put, let us assume that all the samples for training (test, train, validation) are split into k sets (where k > 3). Then, if we shuffle the k sets and assign them to test, train and validation (with correct proportions, train should be about 60% of the samples and test and valida-tion 20% each), we force the model to better generalize its learning capabilities. To make the model more useful, we provide it as a service to the FENIX platform via a REST API. As a matter of fact, all four models documented in this chapter are provided in that way. This isolates each model for potential future extensions, addi-tion of features etc., while they seamlessly work together with the platform. While it is important to document the model's pipeline from training to deploying, it is also necessary to justify its usefulness for a circular economy of reusing materials, especially after going through all this trouble. As already explained, a human expert can do the same as LR does, probably with more precise results. However, there are a few things to consider. For starters, if more variables need to be taken into consid-eration for decision-making, the model can be very easily adapted (to the point of barely changing its code even). In contrast, a human expert will need to adapt his strategy and heuristics, perhaps even do research on new variables, and learn their role and how they affect the result, to achieve similar results. Considering many vari-ables can also lead to mistakes as a single wrong calculation would lead an expert to make the wrong call. The model will never make such mistakes. Furthermore, when facing an industry with ever-increasing needs for fast and efficient processing, it's hard to argue against a fully automated decision support system that can almost instantaneously notify about the quality of the produced material just by taking into consideration the initial parameters it is going to be processed. This knowledge can even be used in artificial experiments to save thousands or perhaps millions of dollars by experimenting with optimal values, instead of trying them and potentially failing.

8.2   Rules Extraction

The next model we are going to present is responsible for extracting rules based on input parameters and pre-classified results. Its function not only complements the LR classifier, it also provides powerful insight on greatly reducing the cost of resources when recycling materials. This will be further expanded upon near the end of the section. The goal of this model is simple: given a set of sample inputs with specified variables and the actual result that is a direct result of these variables, it creates a hierarchy of rules, starting from the most prevailing one and/or selecting the top N rules which guarantee with maximal accuracy a desired result. A careful reader might ask, why maximal and not maximum? Remember that every machine learning model makes mistakes (no matter how small), they merely try to maximize the accuracy they have. This means that the extracted rules will ensure that the model is correct for as many times as it is possible given the data it was trained with. In general, rules extraction consists of a family of powerful inductive algorithms which are based on the principle of separate and conquer. There is a wide area of applications these kinds of models can be applied on: stock investment, finances, text processing and so on. An investor can analyse the global market and use extracted rules to decide when it is wise to invest on a specific stock based on past economic trends. Text processing for a specific purpose can utilize specific rules to decide on thresholds certain heuristics work. Such models are based on decision trees [2, 3] and random forests [4], which generate a hierarchical structure of rules based on their maximal accuracy, then combine them to maximize the overall accuracy. It is important to note that, while other combinations of rules of similar total accuracy may exist, the ones extracted are also the simplest. This is very interesting, because the heuristic criteria needed to fulfil a specific outcome are as few and simple as possible. The model used for the purposes of the project offers a trade-off between the interpretability of a Decision Tree and the modelling power of a Random Forest. Its hybrid architecture consists of both decision and regression trees. After filtering a set of logical rules according to precision and recall thresholds, the higher performance rules are extracted and, after deduplication (it is possible to reach the same rules from different routes), the final set of the best heterogenous rules is generated. The implementation is based on SkopeRules, which in turn is based on the works of RuleFit [5], Slipper [6], LRI [7], MLRules [8]. In our case, the model utilizes the same training samples as the LR model. The difference is that both the input features and the labels are now given; we now want to extract the rules which maximize the accuracy for each label type. So, since "Satisfactory" status can be "Yes", "No" or "Printable", we need to extract 3 sets of rules, one for each label. This time, we split the dataset a bit differently since rule extraction only works for a single label. 3 separate model trainings take place, each model training under sample data for its respective label. A major advantage of this approach is that, once the set of rules for each label is generated, it does not have to be computed again. It can be stored and just returned on demand. As with all models, this is also offered as a separate service via a REST API, allowing future updates or retraining of the model while the FENIX platform is live. As before, we need to discuss the value of the model compared to preferring a more traditional approach, such as employing a human expert. While a human is capable of providing a set of heuristic rules for maximizing the accuracy for a specific label, it is unlikely they will be the simplest possible and it is difficult to provide weights on the importance of the rules' aspects they picked, they can only weigh their importance based on their own experience in the field. On the contrary, automated rule extraction can provide the simplest rules possible given a training dataset, as well as detailed scoring information. We also need to consider the case where more variables need to be considered for decision making. A human expert will need to conduct research on variable thresholds, study their potential in effecting the result etc., not to mention do all the computations again (and running the risk of making a mistake). In the same situation, the model barely needs any change, and, after training, it is ready once more to immediately provide the extracted rule sets. Finally, automated rule extraction can easily enable artificial simulations where power plants can test their results by tweaking the variables around their thresholds. This can lead to cheap discovery of combinations of variables which provide better productivity and with even less cost in resources.

8.3   Time Series Forecasting

Next, we are going to discuss our proposed time series forecasting model. A time series is a series of data points indexed, listed, or graphed, in time order. Most commonly, a time series is a sequence taken at successive equally spaced points in time. Thus, it is a sequence of discrete-time data. Time Series analysis can be useful to see how a given asset, security, or economic variable changes over time. Obviously, any kind of variable evolving through time is an immediate subject of application, such as stock market values, financial values of houses in the 1970s etc. The power of time series analysis lies in taking into consideration more than just the individual discrete data points. It takes into consideration the correlation between two data points, but it also takes into consideration the overall behaviour of the series throughout the entire past to generate the next point. For our purposes, we wish to analyse the active, reactive, and apparent power series in order to predict the future. Since we already have existing samples, we can just get rid of a few samples after a certain time t. Then, we can try to predict them by training on the past points of the series. A very important parameter that greatly affects the capabilities of a time series prediction model is the window size. Simply put, it is the number of discrete data points the series uses internally to infer correlations from. Selecting a large window can lead to overfitting or completely missing the important correlations (due to the model focusing on the general structure of the series instead of specific patterns and correlations). On the other hand, a small window provides little to no correlation information, making the model perform badly. Therefore, it is important to find a window size value which serves as a good compromise between the two. The model we created gets as initial data all existing data points for each reactor variable, then generates incrementally and in real-time future data points indefinitely. It is important to observe here that, since the time series contains numerical values already (we are inspecting a numerical variable's evolution through time), no transformations for input data is required for this model, a far cry when compared to the previous models. As time advances after the initial data points, the series graph is split into two parts: the series evolving with real values and the series evolving with predicted values. This dual graph is provided to assist in decision making, so that any major differences between a predicted and actual value can be immediately spotted. To make the information more accessible, the past of the series is coloured in blue, the predicted series in red and the actual future series in dotted blue. The diagrams also offer dynamic capabilities, such as freezing a part of the series for more thorough inspection, zooming in to take a closer look at a set of data points, etc. A thumbnail of the entire series is also provided to make browsing easier. Time series prediction of this kind is particularly useful for FENIX decision support. Reactors' status does not need to be constantly monitored that way, releasing human resources for other, more immediate tasks. If there is a big difference between a prediction and an actual value, an automatic alert can be immediately forwarded so that necessary actions can be taken on time. Furthermore, estimating the future values of the reactors' variables can make it easier to predict the lifespan of the reactors, their energy needs etc. This can increase savings by a large margin. It also offers the opportunity to maintain the reactors much better by managing their condition and greatly expanding their lifespan. Another strength of this approach lies in its lightweight nature. The initial input sets the limit to how much data will be in memory. After that point, only a window size's worth of data will be in memory since new values are incrementally generated and returned to the FENIX platform via REST. Computations are fast and the memory needs will not increase in the future, making this model even more appealing to utilize. Like all models, this is also provided as an independent service via a REST API, allowing future updates or retraining of the model while the FENIX platform is live.

8.4 Materials Classification

Lastly, we present the materials classifier, which is based on image data. In this case, the input is an image of the material at hand, while the model classifies the image (and, consequently, the material) into one of two categories (good, bad material). Like all previous models, extending the possible output categories to a larger set is trivial. Once more, we first need to delve into the actual model's finer details before showcasing its usage within FENIX. Since the input is images this time around, the first hidden layer of the neural net will be convolutional. This means that, instead of having a different weight for each pixel of the image, only a small set of weights (and therefore a significantly a smaller number of neurons are needed) are being applied to small subsets of the image. The reason this is more promising than a plain neural network is that "local features" found in previous layers rather than pixels are being forwarded and, as a result, the network sees progressively larger and more complete parts of the image. In addition, since neurons focus on learning specific local features instead of learning everything over and over (e.g., degree lines in images, small shapes etc.) for each pixel, they gain a considerable speedup. Also, since the focus now is entire subsets of the image and not pixels, keeping all values can be redundant and therefore it is possible to gain even more in speedup if subsampling takes place.

Layers that do that are called pooling layers and together with the convolutional layers they form the fundamental building blocks of Convolutional Neural Networks (CNNs). Since CNNs have been around for many years and have been used in many applications (e.g., object detection in images, face recognition, OCR etc.), many datasets with images have been readily available by the community to train and evaluate such models. A well-known example is MNIST (Modified National Institute of Standards and Technology database), a large database of handwritten digits that is commonly used for training various image processing systems, with the goal of recognizing handwriting and digits. Another useful source for datasets with tasks like our own is Kaggle. It offers a dataset for an almost identical problem to our own: dog versus cat classification. The dataset contains 25,000 of dogs and cats, offering a generous number of samples to split into test-train-validation. However, for tasks such as material classification and especially when speed is essential in industrial workflows, it might not be possible to wait to obtain a good amount of training samples. While this adds an additional challenge to our approach, there are ways to remedy this. We will discuss two here. The first approach is pre-processing of the initial data and augmenting them. What this means is that we take all available samples and apply random basic yet random image transformations such as zooming, rotations, rescaling etc. If the randomness is uniform enough without any bias, this will generate even more data of a similar nature to the original. This technique can essentially multiply the available training samples. That way, even with very few data, we can produce a full training-ready dataset. The second approach is utilizing a pretrained model on a large dataset and using its bottleneck features for our own model, after fine-tuning the top layers. A pretrained model of this kind has already learned the features which are the most useful. Leveraging these features, we can potentially reach a better accuracy than relying only on available data. A well-known example of such models is VGG16, which is pretrained on the ImageNet dataset.

Before incorporating a model like this into a custom model, fine-tuning its top layers can further improve the results. This is done by instantiating VGG16's convolutional base with weights, adding the custom model on top of it (load its weights as well), then finally freeze the layers of the VGG16 model up to the last convolutional block. The result is a fine-tuned model to the specific task at hand. It is also necessary to illustrate a weakness of utilizing a pretrained model. For a model like this to be useful for general tasks, it must have been trained on very large datasets and, to perform well, it is highly likely to be a model with a complex architecture (this is further evident if you turn your attention back to Fig. 8.2). This means that, to obtain the bottleneck features and fine-tune it, significant computational power is required, while it will not be very fast the first time around. In applications where speed is also an important factor, this might not even be an actual option. While the second approach can potentially produce better results as explained before, we opted to use data augmentation for FENIX's purposes. Since we are addressing a real-life problem involving online decision support and often requiring near-immediate decisions on top of that. Therefore, it is paramount to utilize an approach that can give relatively good results even with few data samples. Data augmentation for our purposes is relatively simple. We split images for materials for which we already

Fig. 8.2 VGG16
architecture

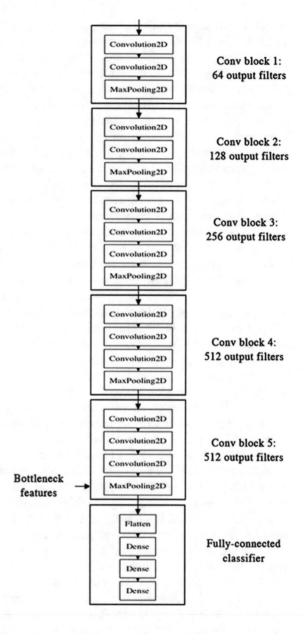

know their class (good, bad) into their respective folders. Then, we augment the data by randomly applying transformations on these images. An image produced by a pre-classified as good or bad will obviously belong to the same class. Taking care to produce an equal number of samples for all classes (two in our case), we further split the data into train, test, validation, splitting each class folder into approximately 60%, 20% and 20%, respectively. Training is relatively fast on a CNN model, especially when leveraging the power of a strong GPU. The result is a model which we offer as a service via a REST API to the FENIX platform. The platform submits a material image and the model almost instantly returns a response saying if it is a good or a bad material.

8.5   Conclusions

Finally, it is interesting to observe that CNN models extract generic features from an image such as lines, objects etc. which help them conduct classifications. While a human expert can do something equivalent, a model can potentially extract features which may not be intuitively responsible for biasing the classification towards one category or another, yet they do. Even better, with a few adjustments, the model can be tweaked to even show how the features it extracted affected its decision-making.

References

1. McCullagh, P., & Nelder, J. A. (1989). *Generalized linear models* (Vol. 37). CRC press.
2. Breiman, L., Friedman, J., Stone, C. J., & Olshen, R. A. (1984). *Classification and regression trees.*
3. Belson, W. (1959). Matching and prediction on the principle of biological classification. *JRSS, Series C, Applied Statistics, 8*(2), 65–75
4. Ho, T. K. (1995). Random decision forests. In *Proceedings of the 3rd International Conference on Document Analysis and Recognition*, Montreal, QC, 14–16 August 1995 (pp. 278–282).
5. Friedman, J. H., & Popescu, B. E. (2005). *Predictive learning via rule ensembles.* Technical Report.
6. Cohen, W. W., & Singer, Y. (1999). A simple, fast, and effective rule learner. In *National Conference on Artificial Intelligence.*
7. Weiss, S. M., & Indurkhya, N. (2000). Lightweight rule induction. In *ICML.*
8. Dembczyński, K., Kotłowski, W., & Słowiński, R. (2008). Maximum likelihood rule ensembles. In *ICML.*

Open Access This chapter is licensed under the terms of the Creative Commons Attribution 4.0 International License (http://creativecommons.org/licenses/by/4.0/), which permits use, sharing, adaptation, distribution and reproduction in any medium or format, as long as you give appropriate credit to the original author(s) and the source, provide a link to the Creative Commons license and indicate if changes were made.

The images or other third party material in this chapter are included in the chapter's Creative Commons license, unless indicated otherwise in a credit line to the material. If material is not included in the chapter's Creative Commons license and your intended use is not permitted by statutory regulation or exceeds the permitted use, you will need to obtain permission directly from the copyright holder.

Chapter 9
User Participation and Social Integration Through ICT Technologies

Aristotelis Spiliotis

Abstract User is one of the most important stakeholder cluster and its participation can link the end of life and early stages in the life cycle of each product when considering the adoption of a circular business model. This chapter presents the main elements of the customer engagement, as identified through a State-of-the-Art analysis carried out in the context of FENIX, as well as those electronic tools in which they will be integrated together with conventional tools for the conduction of commercial activities and the tools to facilitate the interaction with the other actors and activities of FENIX within a single access point digital platform (FENIX Marketplace). The SoA analysis identified the motivational factors that promote a greater customer engagement for the participation throughout all business routes (B2B, B2C but also C2C) applicable in the project. These strategies are improved and enhanced using benefits provided by the social media for the participation in the process. The customer involvement is directly linked to the motives provided within FENIX Marketplace.

9.1 Customer Engagement Strategies

Customer relation has changed over the last years drastically. Initial perceptions view the customer and the brand as discrete entities, a relationship that follows a linear interaction starting from the brand (where value is created) and finishing to the customer (an exogenous factor to the process) [1]. However, literature shows that customers can co-create value, increase competitiveness and be involved in a way to become an intrinsic factor. According to a study by Vargo and Lush [2], as well as another recent by Vivel, Beatty and Hazod [3] on critical elements of engagement strategy, name the following characteristics as essential in engaging the customer effectively:

A. Spiliotis (✉)
Centre for Research and Technology Hellas, Hellenic Institute of Transport, 6th Charilaou-Thermi Road, PO BOX 60361, 57001 Thermi-Thessaloniki, Greece
e-mail: aspiliotis@certh.gr

© The Author(s) 2021

109

P. Rosa and S. Terzi (eds.), *New Business Models for the Reuse of Secondary Resources from WEEEs*, PoliMI SpringerBriefs,
https://doi.org/10.1007/978-3-030-74886-9_9

- Value co-creation: by allowing customers to participate in the process
- Dialogue: which contains the provision for facilitating interaction among all stakeholders and is user to increase the ability of engaging the customers

Which both initiatives are proposed to facilitate actions in specific service relationships rather than direct customer participation. Hence the digital platform of FENIX will consist of 2 main partitions, one to focus into the provision of facilitating interactions and one for the value co-creation. The above strategies are used to engage customers, which in turn, will exhibit a) community and b) transactional engagement behaviours, the first seen as a measure of participation intensity and the latter as a set of activities facilitating repurchase behaviour and strengthening commitment [4]. These behaviours are enabled through relational benefits (see §social, financial) and expressed by recognised types of manifestations which include cooperation, feedback and compliance concerning the B2C interactions but also helping others and spreading positive Word-of-Mouth for the C2C interactions. A general conclusion may include that strategies should be focused on providing the relevant motivational factors (intrinsic or extrinsic) that influence the user behaviour and may lead to specific engagement outcomes, such as satisfaction, loyalty and commitment. It is important to note that no specific benefits lead to engagement behaviours. Some entertainment benefits may lead the user to present either Transactional or Community behaviours.

9.2 FENIX Digital Ecosystem and Provided Incentives

Based on the above, it is concluded that the research model for the digital Marketplace should create an ecosystem where the customer and particular digital tools co-exist along with the provided behavioural benefits (Fig. 9.1).

Customer's manifestations require a mean (a channel) for transmitting the above messages, a platform to state their expression and a network to disseminate them. FENIX digital platform (Marketplace), with its tools, is intended to play a crucial role in all of the above requirements. The network may also be passively enhanced through social media and for that, the most popular channels have been chosen to be linked with FENIX Marketplace (Facebook, LinkedIn and twitter). Social media assists to shift control of some traditional decisions (price, promotion, etc.) to the customers, while at the same time enables customers to participate in strategic choices and co-create value. This is the basis for the creation and maintenance of the Brand Community (BC) that can perform many important actions on behalf of FENIX and hence the digital platform should provide interoperability by means of content sharing with these channels.

Fig. 9.1 Digital ecosystem

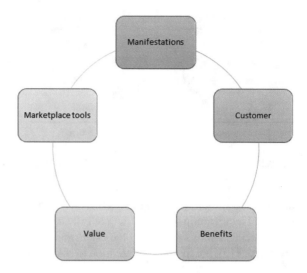

9.2.1   Social Benefits

Social interactions, along with the aid of the social media, helps to swift control of some traditional business activities to the customer (price, promotion, etc.) and allow user to participate in the value co-creation jointly with the brand. Social networking practices are those focusing on creating, enhancing, and sustaining ties among community members, such as welcoming, empathizing and governing [5]. Consumers often also participate in the community to seek assistance and help from other members in supported discussions lined with social conversations [6]. A social platform should be able to provide Peer-to-Peer (P2P) interactions to accommodate sharing of personal experiences, exchange influence and acquiring cognitive competencies [7]. This model may greatly influence the loyalty behavior within the online community, combined with the special ideological characteristics that FENIX attracts but also endorse recycling values, prospect for new sustainable business models, awareness on manufacturing concernment, general information sharing, etc. (Fig. 9.2).

9.2.2   Entertainment Benefits

Entertainment benefits are mainly provided through gamification and refer to adapting a process to have entertainment qualities of a game. Gamification is the systemic addition of game elements into services [8] such as point, badges, quests and leader boards and is distinguished from games as it does not include pure game mechanisms, but its intention is to enable intrinsic motivation associated with entertainment. Entertainment will be achieved by using the 3 core elements that evoke

Bronze	Silver	Gold

Fig. 9.2 Awarded trophies for the first 3 ranked users on the FENIX leaderboards

intrinsic motivation according to a study by Malone [9], Challenges, Curiosity and Fantasy. Gamification in FENIX will only be used as a motivational and engagement tool associated with increases in the extent and quality of effort that users put into a given task [10], but there is no intention to sustain a gaming platform.

9.2.3   Economic Benefits

The virtual world has provided many tools to connect not only companies but also consumers as it is already accepted that online communities allow strengthening consumer relationships and engagement [11]. Dissatisfied customers are confronted with the dilemma of taking their transaction elsewhere which will result either in giving up the economic benefits or continuing repurchasing but accepting lower levels of satisfaction, so without the economic incentives, consumers are less likely to maintain loyalty or engage in the repurchasing behavior [12]. The financial motives utilized in FENIX Marketplace are based mainly on the altruistic motive of Special offers/prices and less in Monetary compensation, as this had the smallest influence on customer's willingness to engage according to a recent study [13].

9.3   FENIX Crowdsourcing System

In a Crowdsourcing System, tasks are distributed to a group of users for carrying them out, hence a successful system should firstly have available a network of people which can interact and share information in direct and fast way to form a virtual online community and secondly to match the demographics and the special characteristics of the available population to the core values of the project. The motivational factors (intrinsic or extrinsic) that influence the performance of the system have already been identified to the appropriate engagement strategies.

Fig. 9.3 FENIX stakeholder clusters

9.3.1   Role of Human Users

Users in the Marketplace are classified into 2 main categories in each of them there may be more subcategories based on the special characteristics (skills and willingness to get involved) of each user: (i) the Task Providers and (ii) the Task Contributors. The Forum presented in this chapter forms the place of the FENIX virtual community in which a community of users will be configured, with the appropriate features and interests to constitute the human resource of the FENIX crowdsourcing system. Within the presented research model, highly engaged customers is an essential part of the circular process in value co-creation as to provide feedback, participate in product design or assembly, trade reused items or feed the production plants with new raw materials from waste. The potential users have been defined from the initial objectives of the project and validated throughout the consortium meetings and workshops. Figure 9.3 shows the applicable clusters, which is by no means restricted to them, but it may rather be expanded to include additional actors in the future.

9.3.2   Crowdvoting System

Voting systems typically require from the requester to select one or more answers from a number of choices (provided by other crowdsourcing workers). Users can upvote, downvote or not vote so that answers and questions can be ranked among the community. Such mechanisms assist to evaluate the quality of data since the contributors are not experts and usually their incentives are not aligned with those of the requesters. Voting is also required to evaluate other behaviours and actions (Fig. 9.4).

Like   Thanks   Laugh   Confused   Sad

Fig. 9.4 FENIX crowdvoting emoticon system

Downvote   Upvote

Fig. 9.5 Two-way voting system

Bad   Poor   Neutral   Good   Excellent

Fig. 9.6 Community reputation recognition

Voting system uses 2 types of regular voting, one for the evaluation of the content (Figure 9.5 which also includes the offered products for sale/ trading). The "Like", "Thanks" and "Laugh" vote gives Positive (+1) reputation, the "Confused" gives Neutral reputation and the "Sad" gives Negative (−1) reputation. On the same way, "Upvote" gives Positive (+1) reputation and "Downvote" gives Negative (−1) reputation for the user profile (Fig. 9.6). A flag warning can also be used to directly report activities that deviate from the regulations.

With respect to the voting user profiles are classified based on their community reputation (qualitative dimension) and their community ranking (deputizes the extend of the user's activity). A newly introduced member is ranked as "Neutral". It is evident the gamification elements that will boost user entertainment through these crowdsourcing activities.

9.4 FENIX Pre-identified Goals and Link to Developed Mechanisms

A number of crowdsourcing mechanisms have been identified by analysing the three key elements of Web-based crowdsourcing [14]: the crowd, the outsourcing model and advanced internet technologies to be used. The demographics and the purpose of crowdsourcing has been studied extensively and below the most important goals of are captured, matched with the most suitable solution approaches, the triggering factors and the platform on which it is intended to function. These functions and the appropriate mechanisms have been developed in the context of WP5 and presented in the next section (Table 9.1).

So, the following list presents the main goals which should be aimed at fulfilment by the functions of the digital Marketplace:

Table 9.1 Main objectives of WP5 coupled with an appropriate CS mechanism

No.	Aim	CS mechanism	Operating platform
1	Populating item availability	Customer generated content platform (upload own designs, items, etc.)	Marketplace
2	Optimizing demand-production	Allow and promote mass orders, DSS link with sensors, etc.	Marketplace/DSS
3	Main and secondary market	Trading platform	Marketplace
4	Product-service evaluation	Enable a crowdvoting system	User forum and Marketplace
5	Minimizing fraud/scam risk	Use of crowdvoting system for reporting inappropriate activities	User forum and marketplace
6	General problem solving	Open content communication tool	User forum
7	Identifying future market trends	Monitor customers activity and feedback, but also allow information fishing through custom polls. Also use data mining in the search bar	User forum—Data mining and Polls, marketplace—Data mining
8	Attract new customers	Allow interactivity with popular social media for dissemination	User forum and marketplace
9	Promote and support the creation of new innovations	Provide a platform where individuals can set custom challenges and call for contributors	Marketplace/open innovation platform

- Populating the availability of new items
- Optimising demand-production
- Main and secondary Market
- Product-service and content evaluation
- Minimising fraud/scan activity
- General problem solving
- Identifying future market trends
- Attract new customers
- Promote and support the creation of new innovations.

9.5 The Digital Marketplace

The Marketplace has been designed to form a very powerful tool which will host the appropriate functions that satisfy and promote both the activities of FENIX, as well as the customer engagement strategies. The main platforms included within are:

- The Marketplace
- A Forum
- A Showroom
- Customer's generated content page
- An Open Innovation Platform (OIP).

which are accessed through a single access point, using a FENIX account or use an existing account from Facebook or Google. Other minor functions that operate horizontally exist and function in addition to the above, but with significant importance. In the following sections, a presentation of each platform and functions is included. Access to the application can be obtained through the weblink: https://forum.fenix-apps.eu/.

9.5.1 Forum

The user forum is intended to form the online social community of FENIX, connects users through Peer-to-Peer interactions and it includes the feedback collection mechanisms. It is associated mainly to the social benefits provided towards customer engagement, but also encompasses entertainment benefits incorporated within horizontally applied functions (e.g., crowdvoting). Users are separated into the following groups that represent a membership which is accompanied by consistent privileges and responsibilities within the online society (administrators, moderators, registered users and visitors). Administrators manage technical details and other core operations of the platform, moderators are granted access to monitor the content of posts and threads, while registered users can post content and participate in online activities as well as to be members of the FENIX social groups. The unregistered users can only view and read content. Social groups are formed to represent the active social clustering (see Sect. 9.3.1) within this ecosystem and their members share very close interests and background and hence enhance the social cohesion among users. This characteristic together with the integration of content exchange with popular social media is expected to expand the competence of the digital ecosystem and increase the Word-of-Mouth activities, which in turn is usually accompanied by other desired crowdsourcing activities such as advertising and initiation. Other groups can be freely created by the users. The Forum section is divided between 2 sub-sections, one which is related to the forum itself and the engaging of users in free text dialogues and another one called Reviews, which is used to capture in quantitative form the feedback of the users (Fig. 9.7).

In the first section users can share any content of their interest with the community, make comments, provide review and develop discussions with other members through open discussion topics and threads and aims to achieve the goals set for the product/service evaluation (Table 9.1, No. 4) and general problem solving (Table 9.1, No. 6). Additionally, in this section partners of FENIX consortium have created threads corresponding to the scientific fields and topics covered by the project to

Fig. 9.7 Forum section overview

inform users about their activity, special thematic information and engage in direct dialogue throughout a "Questions and Answers" section in each topic. Every post in this section can be subjected to voting using the horizontally applied crowdvoting mechanism (Sect. 9.3.2) and share it through the most popular social media for dissemination and word-of-mouth. The Review section utilizes active polls in which users can participate with quantitative input to extract fast and easy-to-use feedback. Through the processing of these results stakeholders of FENIX may easily identify the potential of a product in the market penetration and even produce significant insights about future market trends based on the stated preferences of the users (Fig. 9.8).

Fig. 9.8 Example of a finished review thread. Quantitative results are produced automatically

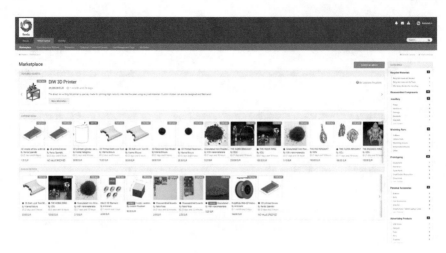

Fig. 9.9 FENIX marketplace main trading portal interface

9.5.2 The Main Marketplace

This serves as the main trading place of FENIX ecosystem (goals set in Sect. 9.4). Trading is strictly performed through financial interaction, but users can also choose to exchange items or set other requirements based on prior mutual consultation. The interface comprises of a list of products either new or second hand and services offered within the FENIX business model. Several filters will be available as an option for the user to apply his/her particular search preferences (product or service, new item or second hand, sorting order by newest, price, popular, etc.) (Fig. 9.9).

When selecting an item, the user is directed to the item page where additional options can initialize horizontal functionalities such as payment methods to proceed to the purchase of the item, start conversation with the seller to engage in negotiations, ask questions about the item, rate it and many more. Also, this platform may be used by the users to upload their own items in order to become active members of the business model by selling/trading old items for recycling or to create secondhand items for sell and thus extend the life cycle of a product or even link the end of life to the early stages of new products (recycled items will provide recycled raw material for the construction of new throughout FENIX use cases).

9.5.3 Showroom

In the Showroom, all the evaluated services, products, content and activity are displayed in ranking order, the best of which occupy prominent and distinguished positions on the screen. This is mainly an entertainment element which promotes

Fig. 9.10 Member contribution leaderboards

users into a productive contest among them, but also serves the social incentives through recognition for their contribution to the FENIX community (Fig. 9.10).

The above ranking makes evident the user's bond with the online community. Three different leaderboards are produced regularly to cover activity related to: (a) Reputation, (b) Content created and (c) Most posts, while virtual trophies (gold, silver and bronze) are awarded for the first positions.

9.5.4 Customer's Generated Content

In this platform all the uploaded content by the users will be available so that member's content can be easily retrieved from the respective database. It is a digital Marketplace with similar interface to the main Marketplace (Sect. 9.5.2), with the difference being that tradable content in this section is linked to the individual effort of the users towards value co-creation. Materials or immaterial work or services are solely user generated and participation has a direct financial incentive as designs, for example, or other services can be purchased in return for a financial profit (Fig. 9.11).

9.5.5 Open Innovation Platform

FENIX Open Innovation Platform (OIP) is the virtual place where the needs of the various stakeholders can meet, and match challenges defined by Task Providers to a pool of Task Contributors. Within this platform any user can set up a new task by defining the title together with the description of the requested input, upload related material and fill in the available timeframe for submitting proposals (Fig. 9.12).

The contributors will be awarded a price based on social, financial or entertainment incentives. This platform will ensure that new innovations will continue to take place by utilizing crowdsourcing through rewards and FENIX platform will provide a direct link to the business chain.

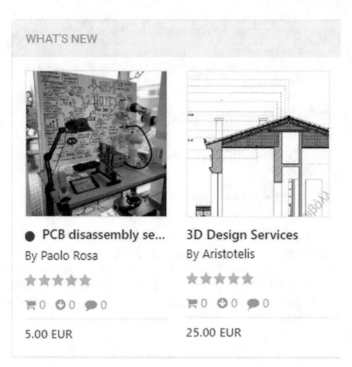

Fig. 9.11 Advertisements on the user generated content section

9.5.6   Other Horizontally Applied Functions

Some functions are applied horizontally, which means that can be found in 2 or more sections/ parts of the Marketplace. One such function is the algorithm that scans the content within the whole platform, to identify activity related to fraud or inappropriate content. Of course, this goal is also served and enhanced by the activity of the Moderators and Admins of the platform. In addition, crowdsourcing is utilized through the crowdvoting system or by a special review system that uses flags. Issuing of a flag leads to the direct examination by the admins of the warning, who then take actions to ensure the smooth operation of the application. Very low evaluation points also result in the examination by the admins for possible inappropriate activity, e.g., task provider does not rank contributors ethically to avoid the fee payment. Another function is the data mining mechanism. The intention of this application is to capture, anonymously and in accordance with the GDPR rules, what the users are searching in the search box. This will produce a database where useful conclusions can be drawn for the user needs and the market trends.

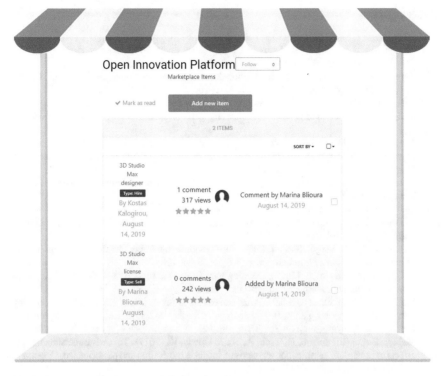

Fig. 9.12 FENIX open innovation platform interface

9.6 Conclusions

A State-of-the-Art analysis was carried out to identify the key enabling mechanisms of Customer Engagement associated with the behavioural manifestations based on the types of the perceived benefits (social, entertainment, economic). Having defined the appropriate crowdsourcing elements for the respective circular business model, as well as the respective incentives, a powerful Marketplace tool was presented which integrates those functions. The User Forum part of this tool enables the formation of an online community by supporting free text content to generate social activity and hence strengthen bonds between members; a poll section will be used to capture the quantitative feedback from the users. On the other hand, the main part of the Marketplace (called again Marketplace) aims to support the economic interactions between the different stakeholders (B2B, B2C and C2C) and host the most important crowdsourcing elements presented in this paper. Those have been framed by ancillary applications to enhance the effectiveness towards the implementation of a truly innovative cyclical economy business model.

References

1. Deshpande, R. (1983). "Paradigms lost": On theory and method in research in marketing. *Journal of marketing, 47*(4), 101–110.
2. Vargo, S. L., & Lusch, R. F. (2008). Service-dominant logic: continuing the evolution. *Journal of the Academy of marketing Science, 36*(1), 1–10.
3. Vivek, S. D., Beatty, S. E., & Hazod, M. (2018). If you build it right, they will engage: A study of antecedent conditions of customer engagement. In *Customer engagement marketing* (pp. 31–40).
4. Dovaliene, A., Masiulyte, A., & Piligrimiene, Z. (2015). The relations between customer engagement, perceived value and satisfaction: the case of mobile applications. *Social and Behavioral Sciences, 213,* 659–664.
5. Schau, H. J., Muñiz, A. M., & Arnould, E. J. (2009). How brand community practices create value. *Journal of marketing, 73*(5), 30–51.
6. Dholakia, U. M., Blazevic, V., Wiertz, C., & Algesheimer, R. (2009). Communal service delivery: How customers benefit from participation in firm-hosted virtual P3 communities. *SSRN Electronic Journal, 12*(2), 208–226.
7. Lusch, R. F., & Vargo, S. L. (2006). Service-dominant logic: reactions, reflections and refinements. *Journal of Marketing theory, 6*(3), 281–288.
8. Huotari, K., & Hamari, J. (2012). Defining gamification: a service marketing perspective. In 16th international academic MindTrek conference.
9. Malone, T. W. (1980, September). What makes things fun to learn? Heuristics for designing instructional computer games. In Proceedings of the 3rd ACM SIGSMALL symposium and the first SIGPC symposium on Small systems, Palo Alto, California (pp. 162–169).
10. Mekler, E. D., Brühlmann, F., Tuch, A. N., & Opwis, K. (2017). Towards understanding the effects of individual gamification elements on intrinsic motivation and performance. *Computers in Human Behavior, 71,* 525–534.
11. Algesheimer, R., Dholakia, U. M., & Herrmann, A. (2005). The social influence of brand community: Evidence from European car clubs. *Journal of marketing, 69*(3), 19–34.
12. Zheng, X., Xiang, L., Liu, I., & Zhang, H. (2012). Enhancing consumer engagement in online shopping platforms through economic incentives. In Proceedings of the Eighteenth Americas Conference on Information Systems. AMCIS 2012 Proceedings, Seattle, Washington.
13. Fernandes, T., & Remelhe, P. (2016). How to engage customers in co-creation: customers' motivations for collaborative innovation. *Journal of Strategic Marketing,* 311–326.
14. Saxton, G. D., Oh, O., & Kishore, R. (2013). Rules of crowdsourcing: Models, issues, and systems of control. *Information Systems Management,* 2–20.

Open Access This chapter is licensed under the terms of the Creative Commons Attribution 4.0 International License (http://creativecommons.org/licenses/by/4.0/), which permits use, sharing, adaptation, distribution and reproduction in any medium or format, as long as you give appropriate credit to the original author(s) and the source, provide a link to the Creative Commons license and indicate if changes were made.

The images or other third party material in this chapter are included in the chapter's Creative Commons license, unless indicated otherwise in a credit line to the material. If material is not included in the chapter's Creative Commons license and your intended use is not permitted by statutory regulation or exceeds the permitted use, you will need to obtain permission directly from the copyright holder.

Chapter 10
Recycling and Upcycling: FENIX Validation on Three Use Cases

Alvise Bianchin and George Smyrnakis

Abstract Within FENIX a set of three use cases has been developed in order to test in practice the selected business models. After the description of a common data repository virtually connecting all the use cases and describing the common step of PCB disassembly, this chapter presents each use case into detail, by evidencing the main findings.

10.1 Introduction

FENIX links together different production steps with the final goal to effectively re-introduce in the market the material recycled from wastes if electrical and electronic equipment. This chapter describes the work done for the technological validation of the pilot manufacturing lines networked together, to validate the links between them: assembly/disassembly process, hydrometallurgical process, High Energy Ball Milling (HEBM), feedstock formulation and 3D additive manufacturing. Different material has been tested also to identify the most effective cooperation and the results applied to support the definition of new business models. Three use cases have been considered for the validation of the net-worked activities of the plant: the direct use of the recovered PM by additive manufacturing (Use Case 2), the valorization of the by-product of PM recovery in form of metal powders for Additive Manufacturing via Robocasting and in form filament for Additive Manufacturing via Fused Filament Fabrication. For each use case several iterations of between the networked plant have been performed to identify optimization points and to collect da-ta on the networked

A. Bianchin
MBN, nanomaterialia Spa, Via Bortolan 42, 31050 Vascon di Carbonera, Treviso, Italy

G. Smyrnakis (✉)
I3DU, Australia Avenue 114, 85100 Rhodos, Greece
e-mail: george.smyrnakis@newcastle.ac.uk

© The Author(s) 2021
P. Rosa and S. Terzi (eds.), *New Business Models for the Reuse of Secondary Resources from WEEEs*, PoliMI SpringerBriefs,
https://doi.org/10.1007/978-3-030-74886-9_10

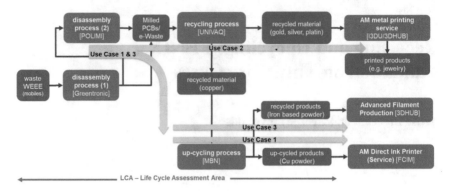

Fig. 10.1 Process chain for the three FENIX use cases

plant that can be utilized for Business Development, Life Cycle Performance Assessment, and circular supply chain digitalization. The combination of the three use cases offers also interesting datapoints to evaluate the efficacy of a circular business based on multiple streams of activities (Fig. 10.1).

10.2 Albus, the Data Repository of FENIX

The Use cases have defined a unique repository for the data generated during the validation, and in particular characterizing the batches delivered within the consortium. This platform, named internally Al-bus, consists in a space online in which pictures, analysis, processing data, identification codes are uploaded. Albus is the primary source of data for the validation of the business cases and the evaluation of the benefits of the Decision Support System to the whole business cases. The batch identification system that has been put in place allows to track the evolution of a component, even a specific phone from grave to cradle as a new product. Each step in the value chain has a defined input and output batch, as well as a defined set of information that characterize the batch. For example, the batches of dismantled WEEE components are provided with information about WEEE class, Component type, Nominal value €/kg, Quantities but also processing time and effort. The adoption of Albus gave the possibility to create a Data flow along with the Material flow that helped in the understanding and quantification of the business opportunity (Fig. 10.2).

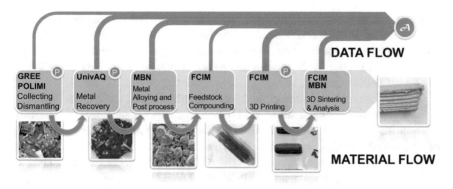

Fig. 10.2   Material and data flows in FENIX

10.3   Collection and Dismantling, the Conventional Approach

The initial steps in the value chain are similar, regardless the down-stream activities afterward. The first step in the networked pilot plant consists in the collection of WEEE to be recycled. For the first set of tests performed on the pi-lot plant, GREENTRONICS collected a quantity of 12.2 tons of electrical and electronic waste (WEEE), PCs and laptops mainly. The WEEE entered the normal treatment flow, respectively manual dismantling and sorting by types of fractions (electronic boards, iron, aluminum, coolers, power supplies, plastics, etc.). The fraction that was of interest in the first set of tests was the waste printed circuit board (WPCB), category 1. Thus, after the first treatment phase resulted in a quantity of 1 ton of WPCB category 1A. In the second treatment phase, the WPCBs were depolluted according to the UNIVAQ specifications. The following components were manually removed: batteries, electronic capacitors, heat sinks, connectors, quartz resonators, inductors, black panels, and multilayer and Ta capacitors. In the end, a quantity of 502 kg of cleaned WPCB resulted. In the last treatment phase, the WPCBs were ground in a professional shredding machine in 0.50–3.00 cm particles (Fig. 10.3).

Fig. 10.3   WEEE collected and shredded for recycling at the Hydrometallurgical Plant

Other batches for recycling tests have been collected, varying com-ponents existing on the WPCB: components rich in rare earths and rich in precious metals. GREENTRONICS collected the same amount of WEEE and went through the same treatment phases, except for the shredding part of WPCB. A quantity of 43,735 kg of cleaned WPCB resulted.

To fully assess the potential of the hydrometallurgical plant, and the downstream metal valorization, components with a high content of precious metals (gold, silver) has been also collected. GREENTRONICS proceeded to a more detailed dismantling of the motherboards, extracting from them the RAM memories and CPUs. To collect 25 kg of such selected components, GREENTRONICS processed an amount of 1060 t of WEEE. In the first stage WEEE was manually disassembled, in fractions (electronic boards, iron, aluminum, coolers, power supplies, plastics, etc.), and then, in the second stage, we manually extracted CPUs and RAMs from the electronic boards. Detailed disassembly reports have been collected and uploaded into the platform Albus, to facilitate the evaluation of the LCPA and business figures.

10.4 Semi-Automated Disassembly, an Innovative Approach

A semi-automated PCB disassembly station, set up by POLIMI, was used as platform to evaluate the impact of mechanized disassembly in the business scenarios of FENIX and as glimpse of what could be a fully mechanized system form the point of view of data analysis, process optimization and business digitalization. Considering both the same process and the same equipment, a dataset has been implemented based on disassembly tests done on more than 50 PCBs previously extracted from WEEEs by GREEN operators and sent to POLIMI for a further disassembly. These PCBs came from disposed cellular phones. The "Albus" platform has been populated with several pieces of information to create a logical (and continuous) information flow among the different pilots. The data gathered by POLIMI do not have a practical value because of the low performance of the presented process compared with existing industry-oriented ones. Instead, it must be intended as a tentative to link the developed PCB disassembly process with the other pilots (Fig. 10.4).

The data gathered during the PCB desoldering process can be divided into two parts: data gathered from the cobot and data gathered from the operator. The need for two different data sources comes from the difference in data origin. The data coming from the operator are related to the physical properties and characteristics of the PCB while the data generated by the cobot are related to the automated process. Further sources of data have been considered to increase the efficiency of the data gathering process:

– The introduction of a digital load cell under the PCB fixture to en-able automatic weight measurement of the PCB before and after desoldering.

Fig. 10.4 A semi-automated disassembly station

- The addition of a power measurement system capable of logging data directly into MQTT.
- The use of real time data to automatically identify the phase of the process being performed and tag the data with this information to add more knowledge on the process and on its sustainability.
- The exploitation of a cobot-assisted process just for the desoldering of both valuable and already functioning components from PCBs that could be, eventually, re-sold into the market as secondary spare parts.

10.5   Use Case 1: Green Metal Powders for Additive Manufacturing

The new Robocasting machine has been designed to use feedstock composed of advanced metal powder formed with non-precious metals recovered from electronic waste. The machine can manufacture metal parts by additive and this enables FENIX to target a new segment of the AM market which has substantial growth potential over the next few years. This use case involves most of the plants developed in FENIX: Electronic wastes are treated via a Hydrometallurgical process to ex-tract precious metals, in which the resulting leachate is still rich in metals such as Copper that is collected and valorised. The copper is collected by electrowinning and mechanically alloyed with primary metals to form a new metal alloy. This powder is com-pounded with hydrogels to form an ink and is packed in syringes to be used by the dedicated DIW machine. The part realized with this AM technique can be afterward treated in a furnace to remove the hydro-gel additives in a first step and to sinter the part in a solid metal afterward (Fig. 10.5).

The setup of the different production units and the parameters of the process involved have been adjusted to the specific requirements of this use case, from step 3 onward. From the interactions between the production steps useful information have been retrieved on the aspects of the business, both from the technical point, i.e. most

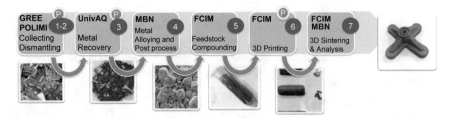

Fig. 10.5 Material evolution between the different processing steps

effective parameters and procedures, and the commercial point, i.e. manufacturing lead time, costs etc. For this use case the main hydrometallurgical process utilized has been the GoldRec1, the plant was built to work following multiple pathways and differentiating between the use cases was useful to test the re-configurability of the networked plants (Fig. 10.6).

As explained in Chap. 5, the GoldRec1 final step consist in the electrowinning process, and this determine the specific challenge for USE CASE 1: the optimization of a Copper resulting from electrowinning of the residual solution from the recovery of the much valuable Au and Ag. The metal obtained as it is, find scarce if no use due the high oxygen level and the heterogeneous morphology of the particles, that makes it not competitive in the market if not further processed by casting or atomization. MBN applied its mechanical alloying route to this upcycling task, processing the copper to reduce the oxide content and integrating it in a new alloy suitable for additive manufacturing processes that involves a final sintering step (Fig. 10.7).

Fig. 10.6 FENIX hydrometallurgical plant: 3D view, in red highlighted the components used both in Gold-Rec 1 and 2, in green the equipment for electrowinning used exclusively for GoldRec1

Fig. 10.7 Upcycling steps for copper powder

Fig. 10.8 As received copper—Obtained from electrodeposition

In step 1, to provide consistent data on the powder characterization, the powder had to be homogenized, this step is anyway required to make the powder suitable for the mechanical alloying process with the MBN's plant. It is important that the powder resizing does not substantially affect the material properties, such as composition, microstructure, and homogeneity of phases distribution. Different process energy levels were therefore tested before considering acceptable the resulting powder (Fig. 10.8).

Step 2 is crucial step before processing of recycled Copper is to characterize it, to assess its actual metal composition and to quantify the presence of copper oxides. Visual inspection can reveal that the recycled Copper presents a wide morphological range, that is directly related to the electrowinning process. The copper adhering directly to the cathode appears in the form of scales and foils also in the centimeter range, while copper deposited on top of this layer resulted in a fine powder, once scraped from the cathode (Fig. 10.9).

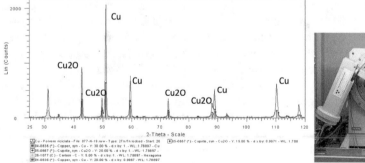

Fig. 10.9 XRD analysis of the recycled copper

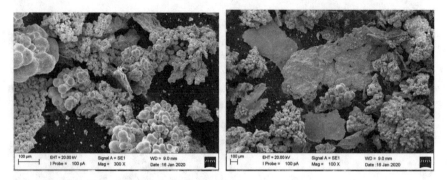

Fig. 10.10 Different morphologies of the recycled copper from electrowinning

For a quantitative assessment of the composition, MBN deployed different techniques, in particular SEM/EDX to identify the actual composition in metals, LECO analysis to precisely quantify Oxygen, and Carbon trapped in the powder and XRF that is complementary to but have advantages in the sample preparation and response speed (Fig. 10.10).

In step 3 a reduction of Copper powder is required to decrease further the presence of oxygen in the recycled powder. The amount of oxygen can be optimized in the recycling step and during electrowinning, but it is difficult to eliminate. The key to re-move the oxygen is to realize the process in an Ar/H2 atmosphere, the powder is put into a furnace and kept at moderate temperatures, to prevent sintering of the powder °C for time ranging from 120 to 180 min. The powder obtained has been further analyzed to assess oxygen evolution and determine the process condition in function of the oxy-gen level.

In step 4 Copper have been alloyed with other elements to obtain a new metal powder to be used in additive manufacturing. Mechanical Alloying (MA) is a solid-state powder processing technique that involves repeated cold welding, fracturing, and re-welding of powder particles in a high-energy ball mill. Mechanical Alloying has been used also to exploit the possibility to adapt to batch-to-batch variations, possibility due to different WEEE batches, and the possibility to substitute Copper in multiple alloys, that might better tolerate a not completely purified copper powder. The use of recycled copper required to develop new processing parameters and methods to improve the morphological characteristics of the material, especially to make it suitable to produce inks for Robocasting (Figs. 10.11 and 10.12).

The use of recycled Copper has been assessed to evaluate the im-pact of the oxygen content in the final sintering behavior of the pow-der. To this end free sintering trials on the powder, obtained with different conditions, have been performed resulting in a good behavior of the sintering also substituting completely the copper in the formulation with recycled copper. Moreover, the proper use of post processing techniques to select the right particle distribution allowed to obtain densities in the sintered bulk that are comparable with those on a reference powder obtained from pristine metal powder.

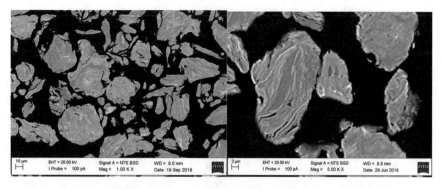

Fig. 10.11 SEM images in backscattering of section of alloyed powder

Fig. 10.12 Sintering trials from powder with different morphologies

The results obtained with the use of recycled Copper from electrowinning encouraged testing its optimization also as pure copper powder. To produce pure copper powder, a dedicated processing equipment have been used. Pre-conditioning of the equipment for mechanical alloying has been done using pure commercial copper obtaining a homogeneous lining that is continuously renovated and maintained during the process. Recycled Copper powder from electrowinning has been processed with the use of a specific Process Control Agent—PCA, to de-crease further the oxygen content. The PCA residues were afterward completely removed from the powder with a thermal treatment at low temperature (300 °C) obtaining acceptable values for residual oxygen levels (<0.5%) although the appearance of the powder resulted in affected and evidently darkened. If the AM processing aims at the commercialization of this powder, then this shortcoming must be considered (Fig. 10.13).

In step 5, the Particle Size Distribution (PSD) can be only partially con-trolled during the mechanical alloying process by using PCA and the specific process conditions. Therefore, in order to have a narrower PSD, sieves or other classification methods have to be used. After the synthesis step, powder is transferred to an air classifier to select those powder fractions (usually fine particles) that cannot be sieved. For dry materials of 100-mesh and smaller, air classification provides the most effective and efficient means for separating a product from the feed stream, for

Fig. 10.13 Post processed powder delivered for ink production and Robocasting

de-dusting, or, when used in conjunction with grinding equipment, for increasing productivity. Air classifiers can only be used for dry processing (Fig. 10.14).

With the alloyed metal powder successfully obtained and manufactured, the next step is to formulate the appropriate composition. The formulation has been done to develop a material ink to be used through Direct Ink Writing (DIW) process. As two different Fe-based powders have been received (monomodal and bimodal), two different types of inks are obtained from the same inf formulation process. The ink must present a pseudoplastic behavior to be printable by DIW. So, it is recommended to have a solid load, i.e. metal powder content, of 35–60% to obtain a functional final

Fig. 10.14 Bimodal particle size distribution obtained with post processing

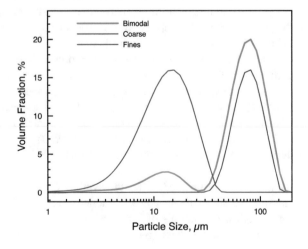

Fig. 10.15 Syringe cartridge with FENIX metal ink being loaded into the shell process

part. The process is based on the mixture of the Fe-based powder with a hydrogel and a dispersant agent. The best performing ink formulation has been obtained with 25% of Pluronic acid as hydrogel and 45% by weight of powder (either monomodal or bimodal) Fe-based Monomodal powder. Dolapix PC75 has been used as dispersant agent (Fig. 10.15).

After loading the cartridge and creating the Digital File with the printing parameters, it is time to start the DIW printing process in the FENIX machine. As stated in previous chapters, the need for the construction of a specific new DIW printer was necessary to cope with the high viscosity of the metal ink. Many printing parameters have been adjusted to obtain good depos-its with the ink created. The main one parameter to be adjusted was the printing speed that has been reduced to improve the extrusion deposition and to get more accuracy in the geometry printed. Temperature also has an important role, the bed temperature was monitored and adjusted according to the ambient conditions. This gave the possibility to have a good improvement in the ink flowability and adherence during the printing. All these improvements have been tested on simple geometries be-fore evaluating the final printing capability on complex and more functional objects (see Fig. 10.16).

The last processing step to be evaluated to before obtaining a full metallic component is sintering the printed part, that is still in its "green" state. A selected number of printed samples in green state have been sintered and analyzed systematically to identify the better performing material and the most suitable/effective conditions that can be suggested to possible users of this technology. Once obtained sintered parts, their visual analysis with SEM is a powerful, although only indicative, evaluation of the actual densification state. To obtain a more quantitative value, that can be utilized also by the DSS, the detailed evaluation of the density has been performed on all the samples. This requires evaluating the picnometric density, in helium, that

Fig. 10.16 DIW process of a complex shape

considers all the open porosities, and the Archimedean density, in water, that take into account only partially the surface roughness but not the open porosities. The internal porosity is calculated considering the theoretical value of a completely densified material (Fig. 10.17).

10.6   Use Case 2: 3D Printed Jewels

Use case 2 focuses on developing a new business model based on an innovative system for offering customized and personalized jewellery to customers using recycled precious metals from the FENIX waste electronics recycling process. A use case pilot is developed, which involves a prototype 3D scanning hardware and the complementary software for customizing jewels. In order to capture the human face, the FENIX team has created a prototype scanner which uses photogrammetry and controlled lighting to accurately capture the human facial geometry and has also developed an initial set of face jewellery designs in order to convert the captured facial geometry to jewels (Fig. 10.14). The purpose of the prototype 3D scanner is to act as a point of sale taking orders directly from customers and feeding them to the FENIX interconnected operations for further processing (Fig. 10.18).

The business model revolves around the concept of specialized hardware that will act as a point of sale taking orders for customized and personalized jewelry from potential clients (the jewelry is 3D printed in castable resin and then cast with recycled precious metals from electronic waste). FENIX developed innovative and customized hardware and software for use case 2, subtracting detailed 3D scan of human faces, by reproducing a 3D mesh of it. Subsequently, the scanned faces will be transformed into a customizable jewelry object. In addition, a software platform will support the commercial process related to customizable face jewelry objects being

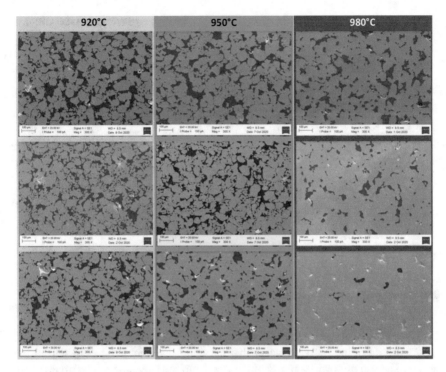

Fig. 10.17 SEM images in section of a sintered 3D printed parts, each row is a material variant—magnification × 100

Fig. 10.18 Face jewelry

ordered from each hardware installation. The 3D printing facility and distribution center that will use the recovered precious metals will be linked to this use case pilot installation and to the online platform that will process all the orders. The process developed is explained graphically in the following Fig. 10.15.

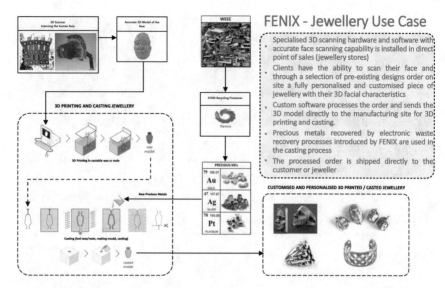

Fig. 10.19 The FENIX customized jewelry process

3D Scanning human faces and converting them into customized jewelry is a brand new and totally innovative concept that currently does not exist anywhere in the world. The use of precious metals recovered from the recycling of electronic waste adds further innovation to the concept and caters for branding the jewels as part of the circular economy. The FENIX Jewelry use case has the potential of a total reconfiguration of the channel landscape allowing jewelers to achieve instant sales with the onsite hardware and the online back end services that will be provided (Fig. 10.19).

10.6.1 Description of the Involved Plants in UC2

Electronic Waste collected and dismantled by GREENTRONIC and POLIMI are processed via Hydrometallurgical route to recover precious metals by UNIVAQ at the plant installed within LoRusso Estrazioni. The precious metal is then melted and converted to jewelry by I3DU and 3DHUB by using lost wax casting in combination with 3D printing (Fig. 10.20).

All the data collected in every step of the production chain are registered and stored in ALBUS, that is the data repository platform for FENIX developed to host the different data sources and to allow linking the batches from WEEE to precious metal.

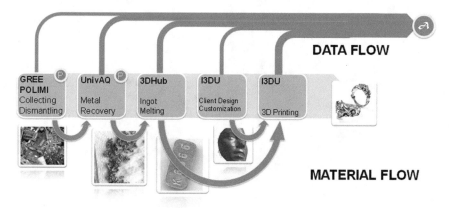

Fig. 10.20 Plants involved in UC2

10.6.2   Metal Recovery

The GOLD-REC 2 hydrometallurgical process (Fig. 10.17), process that was previously developed by UNIVAQ and patented by University of L'Aquila, was applied on grinded RAM modules, PCBs of mobile phones and ceramic CPUs of desktop computer. An important advantage of this process is represented by the fact that it can be performed on the entirely waste material without a preliminary crushing (Fig. 10.21).

The process was tested within the Fenix hydrometallurgical plant (Fig. 10.18) and it consists in solubilization of both base and precious metals content of wastes using the HCl-H2O2-C2H4O2-H2O leaching system (Fig. 10.22).

In brief, the plants involved and their relation to use case two are the following (Table 10.1).

10.6.3   3D Scanning

For the scanning of the human face in 3D and converting the facial characteristics into a usable 3D mesh, several available 3D scanning technologies were originally reviewed. In order to conclude which 3D scanning technology would be the most suitable for the specific application of 3D scanning the human face the following three criteria were considered as the most important:

- **Speed**. A person face is not a static element and is a constantly moving object that cannot hold still for long periods of time waiting to be scanned, so the scan process needs to be either instantaneous or fast.

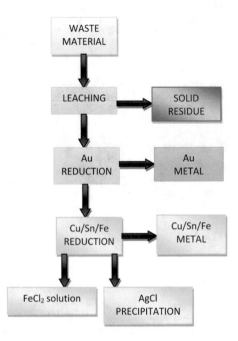

Fig. 10.21 GoldRec2 process flowsheet

Fig. 10.22 FENIX hydrometallurgical plant—3D view

Table 10.1 Use cases relations

Plant/Partner	Action	Relation to UC2
GREENTRONIC	Collecting WEEE	Sourcing of electronic Waste
POLIMI	Disassembling WEEE	Sourcing of electronic Waste
UNIVAQ	Hydrometallurgical processing of WEEE, Precious metal recovery	Sourcing of precious metals
3DHUB/I3DU/JEWELLERY STORES	Development and operation of 3D Scanner	Order taking and processing
3DHUB/I3DU	3D printing + Lost wax casting	Jewelry final product production

Table 10.2 3D scanning features

	Speed	Accuracy	Non-intrusive
Laser triangulation	Medium	High	Low
Structured light	Medium	Medium/High	Medium/Low
Contact based	Low	High	Low
Laser pulse (Time of Flight)	Medium	High	Medium/Low
Photogrammetry	**High**	**Medium/High**	**High**

- **Accuracy**. The resulting 3D scan geometry must be accurate enough so when the jewel is created the human face is still recognisable and the person can identify the facial characteristics on a small scale.
- **Non-Intrusive**. The scan process and underlying technology needs to be non-intrusive for humans.

Of all the reviewed technologies it was concluded that the most suitable one to be used for the application is photogrammetry as it offers the highest speed (click of a button) and is the least intrusive and most familiar one for humans (i.e. it is just cameras photographing the face) (Table 10.2).

For the photogrammetry based scanner that was developed several design variations were designed and several iteration loops were necessary in order to find the optimum angle for the 15 cameras so that there is sufficient camera focal view overlap for photogrammetry to have good results and to be able to create a 3D mesh of the face with sufficient quality for being 3D printed and casted in precious metals (i.e. 15 images of the face taken from different horizontal and vertical angles so that there is more than 20% overlap from image to image).

After the initial prototyping phase, further development of specialized hardware for 3D Scanning of the human face with a more detailed and refined industrial design that looks more like a finished product took place. The most ergonomic and aesthetically nice design was selected as being the best candidate and was further developed into a detailed design. Below are some renders on the final design selected which was the basis for developing the final prototype of the 3D scanner (Fig. 10.23).

That design was also prototyped and constructed using 3D printing (Fig. 10.24).

In parallel with the development of the scanner, development of custom software/firmware for controlling the cameras and picture capture as well as the luminosity from the LED lights took place. The firmware was designed to be able to provide a GUI to the operator enabling them to:

- Control the lighting by adjusting the luminosity of the led lights.
- Control basic image capture properties (brightness, contrast saturation etc.).
- Fully synchronize the image capture so that all 15 images are captured at the exact same instance.

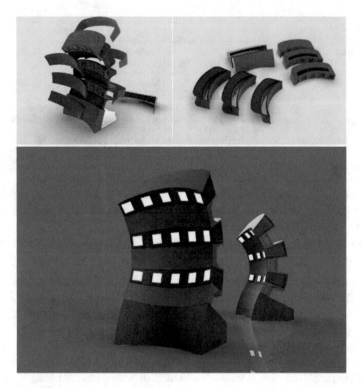

Fig. 10.23 Final detail of 3D scanner design

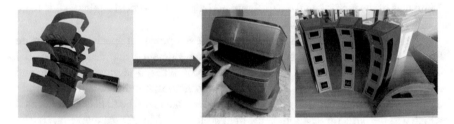

Fig. 10.24 Final 3D scanner design prototype

Apart from developing the firmware a more advanced user interface has also been designed. This user interface will be the main interface that the users will interact with while using the scanner allowing them to 3D scan their face and select and order different custom-made jewelry with their own face in 3D.

10.6.4   Wax Printing and Lost-Wax Casting

Once the human face is captured by the 3D scanner and is fused into the custom jewellery, the production of the jewel can begin. The wax printing and Lost-wax Casting manufacturing process starts by 3D printing the model in wax or castable resin. This step is a type of Stereolithography that uses a wax-like resin. Next, one or more wax sprues will be attached to the model. Then the model will be attached by the sprue to a wax 'tree', together with several other models. The tree is then placed in a flask and covered in a fine plaster. When the plaster solidifies, it forms the mould for casting the metal. The plaster mould is then put in an oven and heated for several hours to a point where the wax is completely burned out.

Next, molten metal is poured in to fill the cavities left by the wax. Once the metal has cooled and solidified, the plaster mould is broken, and the metal models are removed by hand. Finally, the model is filed and sanded to get rid of the sprues. It will be sanded, polished or sandblasted to create the finish desired by the customer (i.e. the cutting-edge technology of 3D printing meets the ancient technique of metal casting!). The upside of casting the model is that the final product will have the best quality: surfaces will be smooth, and the strength, feel, and look of precious metals (e.g. a silver ring or a gold pendant) is exactly what a customer would expect.

10.7   Use Case 3: Advanced Filaments

FENIX linked together different production steps to get advanced filaments for 3D printing using non-precious metals recovered from electronic waste. The metal/polymer filaments can be used in fused filament fabrication (FFF otherwise also known as FDM) to produce metal parts. The development of this feedstock enables the overall FENIX business system to target the prosumer segment of the market which has substantial growth potential over the next years. Like in the other Use Cases, Electronic Wastes are processed via Hydrometallurgical route to remove precious metals, the resulting leachate is still rich in metal such as Copper and Tin that are collected and valorised. The metal mixture obtained after precipitation and calcination is mechanically alloyed with primary metals to form an advanced metal alloy developed to be easily and effectively sintered. The resulting powder is compounded with polymer and additives and extruded in filaments rich in metal (<80 wt%) that can be printed by FFF resulting in a green compound. Debinding and sintering of the resulting part leads to the formation of a 3D printed metal component. The data and results gathered in Albus during the validation of the use case, contributed to the overall assessment of the activities (Fig. 10.25).

This use case is based on the valorization of the by-product form the recovery of Precious Metals with the GoldRec2 approach, these metals consist mainly in Copper Tin and Iron. The details on the processing steps related to the recovery of precious metal are reported in the previous chapter, while the specific part related to UC3

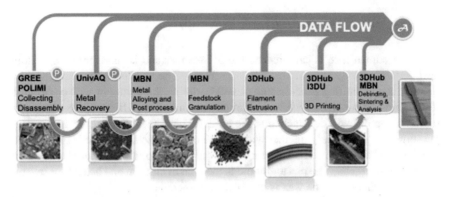

Fig. 10.25 Process and material involved in use case 3

consists in the final cementation of both Cu and Sn with iron metal powder. This powder mixture, as it is, finds scarce if no use due the mixed and variable composition that can also change at each batch of recovery. Mechanical Alloying (MA) was applied in the networked pilot activities to upcycle this metal mix by-product to make it suitable to be used as starting material for other manufacturing routes. During MA, the recycled powder was alloyed with pristine elements to compensate the composition of the recycled metal powder, targeting the production of an alloy suitable for additive manufacturing processes that involve a sintering step. The activities performed in MBN for the powder up-cycling were divided in several steps, going from initial powder analysis to the final granulation that produces a pre-compounded mixture ready to use in AM-feedstock filament production. Already from visual inspection, the recycled powder seams to not suffer much from oxidation, thanks to the calcination step, further analysis confirmed that it can be used directly along with the other commercial raw powder in the MA process. A more detailed analysis at the SEM reveals that the amounts of tiny particles, below few microns, is not negligible, and this might cause problems connected with dustiness, that can hinder a direct commercialization of the recycled powder. The processing steps, as those done in MBN, solve this issue mechanical, by alloying the powder in a new product (Fig. 10.26).

The composition of the recovered powder can vary from batch to batch, according to the hydrometallurgical process parameters, with the composition of the starting WEEE and the efficiency of the process. Therefore, the main initial step in the precise evaluation of the powder mixture is to assess its actual metal composition (Fig. 10.27).

Given the concordance of the data achieved with the two techniques, that is within the respective instrumental errors, and considering that XRF can be directly performed on the powder without sample preparation, this latter was chosen to perform the verification of the batches to verify the metal content and determine the mixture of commercial powder and recycled powder in the mechanical alloying process. The powder must be weighed carefully to respect the weight ratio between the different elements. To improve the alloying process, the powder is mixed to

Fig. 10.26 SEM images of the recycled powder mixture

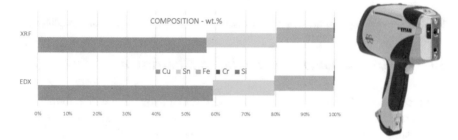

Fig. 10.27 Benchmark of EDX and XRF results on the recycled powder mixture, on the left the portable XRF scanner used for the analysis

start from an already good dispersion of the primary and recycled element. This operation is normally performed also with commercial raw materials, and it is facilitated by the fact that they are purchased all with similar particle size ranges that in general do not contain a significant amount of very small particles. Using the recycled powder metal mix from calcination made this step more challenging, the amount of very fine particles tends to increase the difficulty of an effective dispersion. To overcome this, different approaches have been effectively adopted, one using a dedicated V shape blender, the other using custom made tapered tank for tumbling. Once properly dispersed the powder is introduced in the Mechanical Alloying plant. The processing condition had been optimized to get a dispersion of the elements while maximizing the yield of particles in the fraction between 10 and 60 μm, leaving to the post-processing activities the optimization of the morphology and the yield (Fig. 10.28).

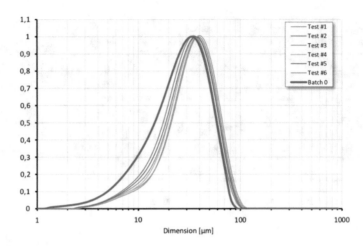

Fig. 10.28 Evolution of particle size distribution obtained with different processing conditions

10.7.1 Granulation, Compounding and Extrusion

The first batches of powder were directly tested for compounding with polymer, but they were not processable because of the high viscosity generated during the mix. Decreasing the metal content in the compound is not a viable solution since it has to be the highest possible, above 80 wt%, to have chances to be successfully sintered in a metal component at the end. Therefore, to overcome this problem MBN deployed its experience in powder agglomeration to provide a pre-compounded product that consist in the alloyed metal powder, mixed with a small amount of binder and additives. Different batches have been produced considering the possible combination of binder, in particular:

- **(M)** Masterbatch powder is formulated to be compounded with more polymer in the extruder, example are the powders with Polyethylene Wax and Stearic Acid for which High density polyethylene (HDPE) is added to make the filament
- **(F)** Production of Final composition has been tried with Acrylic res-in, this was meant to be directly extrudable to make a filament
- **(S)** Granulated powder for Solvent Debinding, where wax is used and must be partially removed with apolar solvents before sintering
- **(W)** Granulated powder for Water debinding, where glycols are used and must be partially removed with hot water before sintering (Fig. 10.29).

The specific extruder utilized to obtain the filament is a "Desktop" grade one. Although the company suggests a maximum filler percentage of 30%, it has been used for processing filled polymers with a filler percentage more than 80%. At this level of filler percentage, the raw materials must be either in powder form or pre-compounded, with the latter being preferable. The ability to process highly field polymers derives from the pineapple—shaped screw. Most of the materials examined were not plastic

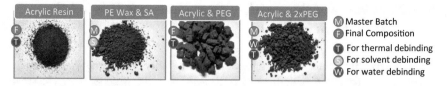

Fig. 10.29   Scheme of the different granulated powder developed for compounding

enough when heat-ed near the melt temperature. That behavior resulted in filament separation when dragged from the puller, restricting the effectiveness of the specific mechanism. In order to solve that issue, the puller's linear velocity should match the filament's one as it leaves the nozzle. Since the installed nozzle had a diameter of 4.0 mm, it was not possible to reduce the filament's diameter if the linear speeds of the pulley and the filament were the same. The solution given was to change the nozzle and install a new one with a diameter of 2.0 mm. That decision was a critical one since it proved to be the game changer for producing high filled filaments with the specific extruder. The most effective combination of metal powder, polymeric binder and lubricant resulted to be the mixture with High Density Polyethylene, Polyethylene Glycol and Stearic Acid (Fig. 10.30).

Once identified the correct extruding parameters the dimensional accuracy was excellent, with the automated extruder function working perfect except when the flow became unstable.

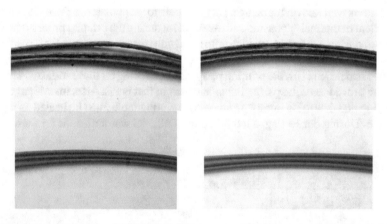

Fig. 10.30   Extrusion trials of HDPE based filament at different screw rotational speeds and temperatures

10.7.2   Printing of the Metal/polymer Filaments Developed from Recycled WEEE

An entry level 3D printer was used to test the printability of the produced filaments. Small square specimens with dimension 20 × 20 × 3 mm were printed. The small size was chosen in order to be easier to investigate debinding and sintering process parameters. Since it was known that a larger nozzle was necessary to avoid clogs, the in-stalled 0.4 mm was replaced with a 0.6 mm one. Although the filaments were fragile, through the developed custom filament holder it was possible to feed them to the printing head without issues. The parame-ters tested were the following:

1. Printing head temperature.
2. Bed adhesion.
3. Bed temperature.
4. Print speed.
5. Infill type.
6. Mass flow.

Printing head temperature is one of the most significant parameters since it affects the liquification of the material, thus the viscosity which in turn affects the material flow. Incorrect temperature setting can cause sporadic clogs and textured top surfaces. The initial testing temperature was 180 °C, but it was concluded that a printing temperature of 186 °C produced the best printing results, as it is presented in the Fig. 10.31.

HDPE is well known for being a very difficult-to-print material since it has a large coefficient of thermal expansion and distorts a lot during the printing process and after printing, during cooling. A good printing bed adhesion is necessary for a successful print. Three substrates were tested, glass plate, PEI sheet and packaging tape. The first two had no success, the printing part was not sticking on these surfaces. The best solution proved to be the packaging tape which in fact is PE. Also, in order to reduce wrapping and distortion during printing, the printing table must be heated as high as possible. During our testing, a temperature of 100 °C was sufficient for these small

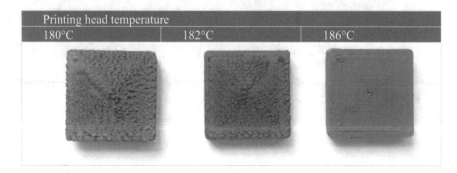

Fig. 10.31 Effect of the printing head temperature on the top surface quality

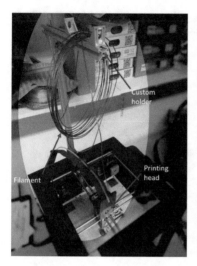

Fig. 10.32 Custom spool/ filament holder to use the filament in a Zortrax M200 3D printer

parts. For larger parts, a warmer setting is advised. Print speed is another parameter that affects the quality of the print-ed object. In the tested case, it was more than obvious that the material itself does not allow fast printings. The initial test speed was 30 mm/s and the optimized one was 13 mm/s. Different printing path were also tested to minimize the presence of voids that would hamper the final sintering step. Although the part with the concentric infill type was heavier, the weight did not match the calculated one considering part dimensions and material composition which should be approximately 3.52 gr. Part porosity is the reason for this issue, and it can be reduced by increasing material flow. After conducting several prints with different material flow coefficients, an increase of about 20% produced acceptable results (Fig. 10.32).

10.7.3 Debinding and Sintering of the Final Metal Part

The setup used for debinding the printed pars is showed in below, a balloon filled with the solvent and containing the samples were kept in a in a thermostated bath for the duration of the debinding. A condenser on top prevents the leaking of fumes, and the whole test has been performed in a fume hood (Fig. 10.33).

The objective of the debinding trials were to:

1. Identify the conditions for a complete removal of the PEG from the sample.
2. Minimise the debinding effort, aiming to ease the process for an end-user.
3. Minimise the risk deriving from using organic solvents (Fig. 10.34).

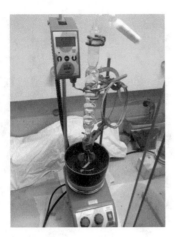

Fig. 10.33 Equipment for solvent debinding

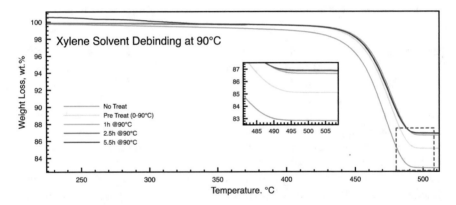

Fig. 10.34 Thermal degradation profile of printed samples debinded in Xylene at 90 °C

Hexane, Cyclohexane and Xylene have been tested, at room temperature and at near to boiling temperature, the best performing process resulted to be with Xylene that also at RT gave better results than the other two. The better performance with Xylene is mostly due with the higher temperature that can be reached in the thermostated bath, thanks to the higher boiling point of Xylene. The sintering cycle has been performed in a tubular furnace, keeping a flow of a gas mixture of Hydrogen and Argon (10% H2) to facilitate binder removal and with the use of Titanium powder as hydrogen Getters. The process is like the one adopted for the sample print-ed via Robocasting in Use Case 1, with the use of alumina sand to prevent distortion of the sample during the thermal process (Fig. 10.35).

On the best sintered samples it was possible to measure the hard-ness, that is indicative of the overall mechanical properties of the specimen. The results are positive and confirming the targeted hard-ness value expected for this alloy. In figure

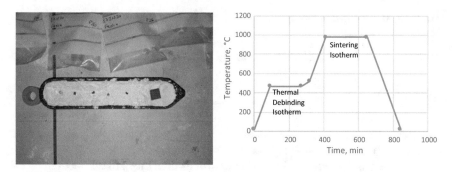

Fig. 10.35 Thermal profile used in the furnace, on the left the sample container with alumina powder

below are also reported the indicative hardness values of Steel and Titanium, where Fenix material can compare. It has been compared with annealed metals since the thermal process, and the following slow cooling, is comparable with an annealing process (Fig. 10.36).

The measurement of other mechanical properties, such as tensile strength and Young Modulus, is strongly affected by the printing parameters and is highly anisotropic.

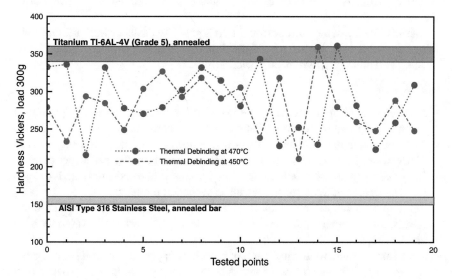

Fig. 10.36 Vickers Hardness measured on different points in the sintered samples, compared with standard materials

10.8   Conclusions

10.8.1   Use Case 1 Conclusions

Affordability

To understand whether a DIW Technology is affordable or not, it must be compared with other processes that could manufacture the same product quality and the best candidate to compare DIW is Selective Laser Melting.

Selective Laser Melting (SLM), also known as direct metal laser melting, is an additive manufacturing technology that uses a power la-ser to melt and fuse powder that is commonly metal material.

Even though SLM has more geometry options with less design limits, both have in common the possibility of adjusting the desired infill and modifying the geometry until it is optimized.

SLM uses layers from 30 to 50 μm thick and DIW layers thickness are between 0.3 and 0.5 mm. In comparison, DIW has the worst surface finish but both need post processing so at the end, this fact is not determinant.

The processes are different, but it is easy to distinguish that DIW has a really simple process compared with SLM, that has to have perfect laser system adjustments.

And finally, DIW has cheaper machine costs. It is around 20.000 and 25.000€ meanwhile SLM is much more expensive and not yet available in the same price target.

The availability of the powder from FENIX processes makes metal printing technology available to a wider range of possible user, that can exploit recycled metal powders as well as standard metal powders.

Main Results:

- If the powder from the recovery plant is meant to be directly sold as a product, a comminution step for homogenization should be considered.
- The connection in a circular business with a powder processing industry such as MBN, waive the need of a powder comminution at the powder recovery plant.
- Recycled Copper after electrowinning require homogenization and oxygen removal treatment to be considered close to market reference.
- The copper obtained from GoldRec1 is almost pure, it does not contain other metals, the characterization procedures can be reliably performed with only XRF and LECO.
- A processing route for the inclusion of recycled copper powder obtained by electrowinning in the new formulation has been defined and utilised to produce Iron based alloy.
- The oxygen level can be reduced but in copper-based powder, solely composed by recycled Copper, this still affect the powder appearance.
- Different solutions have been found to maximise the efficiency of the links between the plants, saving cost and effort in overall production route.

Although the networked pilots can be identified with the production of metal powders, there are more services and products to be considered as exploitable result of the activities, here a list:

- Hydrometallurgical Plant—as product (the plant itself) or as a service for recover of gold and metal powders.
- The powder from recovered metals—as a product to be used for the formulation of inks (also available from FENIX Marketplace).
- The powder from recovered metals—as a product to be used for direct laser sintering in AM (soon available at the FENIX Marketplace).
- Inks and DIW Printers—as a combination of product (he printer) and service (the ink) to be used as affordable additive manufacturing platform to produce parts that can be sintered into metal components (available from FENIX Marketplace).
- Metal Parts AM—as a service, to realize metal parts from CAD design provided by the user (also available in the dedicated section "Customer Generated Content" in FENIX Marketplace).

10.8.2 Use Case 2 Conclusions

The activities performed in the FENIX use case 2 succeed in demonstrating the technical feasibility at relevant industrial scale of the circular business models, by delivering products (raw precious metals) that can be utilized in the jewellery industry. This resulted from iteration throughout all the networked plants, from WEEE collection to jewellery realization, trying different approaches to get the most from the networked pilots.

Although the networked pilots can be identified with the jewellery production, there are more services and products to be considered as exploitable result of the activities, here a list:

- Hydrometallurgical Plant—as product (the plant itself) or as a service for recovery of gold and other precious metals, already working in collaboration with a possible customer, LoRusso Estrazioni
- The 3D Scanner—as a product or as a service, to enable retailers to offer new highly personalized, cutting edge, innovative and green jewellery to their clients.
- The 3D printing facility as a service to jewellers who are selling FENIX jewellery.

10.8.3 Use Case 3 Conclusions

Different solutions have been found to maximize the efficiency of the links between the plants, saving cost and effort in overall production route in view of an integrated circular business. As example:

- XRF characterization agrees with the EDX and is preferred to determine actual composition of each batch of recycled powder before the alloying process.
- The analysis performed in this stage before alloying, that are unavoidable to correctly address the following alloying process, can substitute the characterization step after calcination, saving time/effort in the whole networked pilot.
- The calcinated powder have to be managed carefully for its dustiness, all its handling needs to be performed with correct PPE and/or suction benches.
- The hardware used is scalable, allowing to readily process batches from 20 to 100 kg/day.
- Power post processing has been effective in increasing the overall yield of the powder in the suitable range.
- The Particle Size Distribution has been optimized differently than the material for Use Case One since the granulation and com-pounding process are negatively affected by the increase of surface area associated with bimodal distribution.
- Production method for granulation based on Polyethylene has been chosen among as the most effective one for filament production.

Although the networked pilots can be identified with the filament production, there are more services and products to be considered as exploitable result of the activities, here a list:

- Pre-Compounded powder—as a product to be used for the formulation of filament or for the direct use in injection molding (also available from FENIX Marketplace).
- Metal/Polymer filament—as a product to be used with commercial FFF and FDM additive manufacturing platform for the pro-duction of parts that can be debinded and sintered into metal components (available from FENIX Marketplace).
- Metal Parts AM—as a service, to realize metal parts from CAD design provided by the user.

The results obtained during the activities performed are not limited to the one listed above, the other results, that con not be directly exploited relate to the setup of ALBUS, the development of DSS models for identifying correlations processes in all the networked plants, the mechanical alloying of mixed powders from direct cementation.

Open Access This chapter is licensed under the terms of the Creative Commons Attribution 4.0 International License (http://creativecommons.org/licenses/by/4.0/), which permits use, sharing, adaptation, distribution and reproduction in any medium or format, as long as you give appropriate credit to the original author(s) and the source, provide a link to the Creative Commons license and indicate if changes were made.

The images or other third party material in this chapter are included in the chapter's Creative Commons license, unless indicated otherwise in a credit line to the material. If material is not included in the chapter's Creative Commons license and your intended use is not permitted by statutory regulation or exceeds the permitted use, you will need to obtain permission directly from the copyright holder.

Open Access This chapter is licensed under the terms of the Creative Commons Attribution 4.0 International License (http://creativecommons.org/licenses/by/4.0/), which permits use, sharing, adaptation, distribution and reproduction in any medium or format, as long as you give appropriate credit to the original author(s) and the source, provide a link to the Creative Commons license and indicate if changes were made.

The images or other third party material in this chapter are included in the chapter's Creative Commons license, unless indicated otherwise in a credit line to the material. If material is not included in the chapter's Creative Commons license and your intended use is not permitted by statutory regulation or exceeds the permitted use, you will need to obtain permission directly from the copyright holder.

Printed in the United States
by Baker & Taylor Publisher Services